BestMasters

Alexander Reiter

Time-Optimal Trajectory Planning for Redundant Robots

Joint Space Decomposition for Redundancy Resolution in Non-Linear Optimization

With a Preface by
Univ.-Prof. Dr.-Ing. habil. Andreas Müller

 Springer Vieweg

Alexander Reiter
Linz, Austria

OnlinePlus material to this book can be available on
http://www.springer-vieweg.de/978-3-658-12700-8

BestMasters
ISBN 978-3-658-12700-8   ISBN 978-3-658-12701-5 (eBook)
DOI 10.1007/978-3-658-12701-5

Library of Congress Control Number: 2016930667

Springer Vieweg
© Springer Fachmedien Wiesbaden 2016

Printed on acid-free paper

This Springer Vieweg imprint is published by Springer Nature
The registered company is Springer Fachmedien Wiesbaden GmbH

Foreword

Industrial robotics must address the changing needs of future production systems. Improving productivity and flexibility will remain the major challenge for prospective production systems with reduced energy consumption. The enabling key factors to achieve these goals are novel robotic design principles combined with advanced control concepts. An innovative design principle that is being introduced for industrial robots is the kinematic redundancy, i.e. to use a robotic manipulator that has more degrees of freedom than required to accomplish the intended manipulation tasks. Advanced control concepts shall ensure that a given manipulation task is accomplished efficiently. Time-optimal control schemes are such advanced concepts that allow for task execution in shortest possible time. A seamless combination of these innovative concepts – kinematic redundancy and time-optimal control – does not yet exist.

This master thesis addresses exactly this problem. It presents an original approach to the redundancy resolution that facilitates the numerical solution of the time-optimal control problem. The basis is a non-linear dynamical model for the serial robotic manipulator. Emphasis is always given to a generally applicable approach to the dynamics modeling that allows application of the results to other robotic systems. The presented approach is applicable to any robotic manipulator with a single degree of redundancy, i.e. such that have one more degree of freedom as the task space. The results reported in this thesis represent a progress beyond the state of the art. The reader will get introduced to the basic concepts and to the specific modeling approach.

Optimal control is one of the main research directions of the Institute of Robotics at the Johannes Kepler University Linz, and this master thesis an excellent example for the holistic approach pursued at the institute. In various applications, non-linear, model-predictive, and flatness-based control are also used to derive tailored problem specific solutions. The mathematical basis is always a non-linear dynamical model. This is in particular important for the control of elastic robotic systems. The latter is a research topic that is becoming increasingly important with the advent of light-weight robots. Mobile robotic platforms and humanoids are other topics at the institute that build upon the mathematical modeling.

Andreas Müller
Head of the Institute of Robotics
Johannes Kepler University Linz

Abstract

Industrial robotics applications such as pick-and-place tasks where rapid motions are required, or robot motions along complex paths such as in gluing or laser cutting, have increasingly adopted the use of kinematically redundant serial robots in the last years. Reasons for this can be found in the remarkable characteristics of redundant robots such as the enhanced ability to adapt to the workspace structure or to avoid obstacles by changing the joint configuration of the robot without any end-effector motion. In many cases it is preferred to perform tasks in the shortest possible time leading to time-optimal trajectory planning, the problem of finding trajectories with minimum end times. While solution procedures are readily available for non-redundant manipulators, the challenge of exploiting a robot's kinematic redundancy for minimum-time trajectory planning is not sufficiently covered yet. The present thesis introduces an approach for minimum-time trajectory planning based on a separation method known from literature leading to trajectories that explicitly take advantage of the kinematic redundancy of a manipulator and respect technological and physical constraints of the system. Simulation results demonstrate that the method is applicable to robots of different redundant kinematics and yields time-optimal trajectories.

This work has been partially supported by the Austrian COMET-K2 programm of the Linz Center of Mechatronics (LCM), and was funded by the Austrian federal government and the federal state of Upper Austria.

Kurzfassung

Bei industriellen Anwendungen wie etwa Pick & Place-Aufgaben, bei denen rasche Bewegungen gefordert sind, oder Bewegungsaufgaben entlang von komplexen geometrischen Pfaden wie beim Kleben oder Laserschneiden, werden in den letzten Jahren vermehrt kinematisch redundante, serielle Roboter eingesetzt. Die Gründe dafür sind in den bemerkenswerten Eigenschaften redundanter Roboter zu finden. Dazu zählen die hervorragende Anpassbarkeit der Roboterpose an strukturierte Arbeitsumgebungen und die Fähigkeit Ausweichbewegungen auszuführen, bei denen Robotergelenke ohne Veränderung der Position und Orientierung des Endeffektors bewegt werden. In vielen Fällen ist es gewünscht, Bewegungsaufgaben in möglichst kurzer Zeit auszuführen. Dies führt zum Problem der Trajektorienplanung mit optimaler, das heißt in diesem Fall minimaler, Endzeit. Für nichtredundante Manipulatoren ist diese Thematik bereits weitgehend untersucht während für redundante Systeme viele Problemstellungen ungelöst sind. Die vorliegende Arbeit stellt eine Methode vor, die auf Grundlage eines aus der Literatur bekannten Separationsansatzes die kinematische Redundanz eines seriellen Roboters explizit ausnutzt. Dabei können sowohl technologische, als auch physikalische Beschränkungen des Manipulators berücksichtigt werden. Mittels Simulation wird gezeigt, dass dieses Verfahren für kinematisch redundante serielle Roboter verschiedener Bauarten zeitoptimale Trajektorien liefert.

Diese Arbeit wurde im Rahmen des LCM (Linz Center of Mechatronics) nach dem Kompetenzzentren-Programm K2 durchgeführt und mit Mitteln des Bundes Österreich und des Landes Oberösterreich gefördert.

Contents

List of Figures

List of Tables

1. Introduction

Kinematic redundancy describes a manipulator's topological property of featuring more joints than necessary to assume any configuration in its task space of given dimension. Figure 1.1 illustrates a planar robot with three joints. Since only the horizontal and the vertical position coordinates of the end-effector but not its orientation are selected to be task space coordinates, the manipulator is kinematically redundant. Similarly, the industrial robot from Figure 1.2 consists of a manipulator with six revolute joints on top of a linear axis which also results in the robot being kinematically redundant since the end-effector pose can be described by means of three coordinates each for its position and for its orientation.

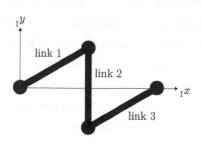

Figure 1.1: Redundant planar robot with three joints, redundancy is introduced by considering the end-effector orientation not a task space coordinate

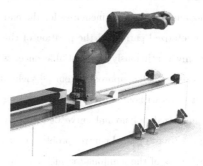

Figure 1.2: Industrial robot with seven joints

1

During the last years, the importance of kinematically redundant serial robots has risen due to striking advantages such as their improved flexibility and adaptiveness in structured workspaces and their inherent ability to perform null space motions resulting in remarkable performance in compliance tasks compared to conventional, non-redundant industrial robots.

The term of *time-optimal trajectory planning* usually, and also in the present thesis, describes the task of finding a minimum-time trajectory, i.e. a trajectory with minimal end time. Introducing time-optimal trajectories to industrial processes may yield economical advantages due to reduced motion cycle times. While for non-redundant serial robots this problem has been comprehensively dealt with, it has not been covered sufficiently for redundant manipulators. Since for redundant robots not only the physical construction, but also the mathematical structure of the resulting physical equations differs greatly, well-known methods for obtaining minimum-time trajectories for non-redundant setups are not applicable to their full extent.

In [9], a method for obtaining minimum-time trajectories along predefined, parameterized task space paths of serial robots with one redundant degree of freedom is presented. The robot joints are divided in a *non-redundant* portion of the same dimension as the task space and a *redundant* part that consists of the remaining joint. First, time-optimal trajectories are obtained for the path parameter and the position of the *redundant* joint by means of an optimal control problem. Finally, the *non-redundant* joint positions are found using analytic inverse kinematics for the end-effector position computed by means of the parameterized paths and the position of the *redundant* joint. The resulting trajectories are only continuously differentiable once, which limits the method's range of possible applications. The approach from [3] yields time-optimal joint trajectories for point-to-point tasks without a predefined connecting path. For each joint trajectory, a separate parametrized polynomial curve of arbitrary degree is assumed, whose parameters are subject to an optimal control problem that also incorporates technological and physical constraints of the manipulator such as joint torques, jerks, accelerations, velocities and positions. Both approaches explicitly make use of the kinematic redundancy of a serial manipulator. However, none of the methods provides trajectories for predefined task space paths with an arbitrary level of continuous differentiability with respect to time while keeping the computational cost at a minimum.

In Chapter 2 of the present thesis, the theoretical background of NURBS curves is presented. NURBS curves are a special type of mathematical curves with properties suitable for trajectory optimization tasks. Chapter 3 outlines the *Projection Equation* from [2], a synthetic method for obtaining the equations of motion of a dynamic system such as a robot. The *Projection Equation* is increasingly advantageous in cases where manipulators consist of a series of similar subsystems such as the planar robot from Figure 1.1 or the industrial robot depicted in Figure 1.2. Chapter 4 discusses the general problem of minimum-time trajectory planning along known geometric paths for kinematically redundant serial robots. Applying the knowledge of NURBS curves for the parameterization of the geometric path, a number of methods that are based on numerical inverse kinematics approaches will be investigated. Additionally, geometric methods to augment the resulting solutions in order to improve specific instantaneous properties of the robot system will be utilized. Chapter 5 introduces a separation approach based on [9] where the kinematic redundancy of the robot is directly exploited by NURBS parameterizations of the path parameter and the trajectory of the *redundant* coordinate. It will be examined whether or not the methods presented earlier are suitable for the given tasks. Finally, Chapter 6 provides two exemplary applications for the optimization formulations presented in Chapters 4 and 5. It will be shown that especially the separation approach presented in Chapter 5 is successfully applicable in many cases. The results of the two examples from Chapter 6 are collected in appendices published as a complementary Springer Online Plus download.

2. NURBS Curves

This chapter provides an overview of the basic properties of NURBS curves and possible applications in the field of optimization in robotics. NURBS is an acronym that stands for *Non-Uniform Rational B-splines*, a class of parameterized geometric curves. For the sake of brevity, proofs are omitted in this chapter and the reader is referred to special literature such as [10].

2.1 Basics

A (in general multi-dimensional) NURBS curve $\mathbf{p}(t)$ of maximum degree n can be represented as a function of the parameter t,

$$\mathbf{p}(t) = \sum_{j=0}^{m-n-2} \mathbf{d}_j R_j^d(t), \quad t \in [a, b], \tag{2.1}$$

defined on m monotonically increasing knots t_k with

$$a = t_0 \leq t_1 \leq \dots \leq t_{m-2} \leq t_{m-1} = b.$$

In general, see [7], the knots are chosen such that

$$a = t_0 = \dots = t_d \leq t_{d+1} \leq \dots \leq t_{m-d-2} \leq t_{m-d-1} = \dots = t_{m-1} = b. \tag{2.2}$$

The curve is called *uniform* if the interval sizes between knots (with the exception of any intervals with size zero at a or b; sometimes also referred to as *open-uniform*) are equal, otherwise it is called *non-uniform*. In (2.1) on the previous page $\mathbf{d}_j$ denotes the control points of the curve that form a polygon, the control polygon, in whose convex hull the curve lies, see the convex hull property in Section 2.1.1. $R_j^d(t)$ is a polynomial piecewise B-spline basis function of degree d that is defined by the rational expression

$$R_j^d(t) = \frac{w_j N_j^d(t)}{\sum\limits_{i=0}^{m-n-2} w_i N_i^d(t)}. \tag{2.3}$$

$R_j^d(t)$ allows to change the influence of each control point $\mathbf{d}_i$ using weight factors w_i and $N_i^d(t)$ are polynomial functions that are non-zero on the interval $[t_i, t_{i+d+1})$, see Section 2.1.1. For the computation of $N_j^d(t)$ see Section 2.1.2.

Rewriting the sum in (2.1) on the previous page, a matrix representation can be found which is advantageous especially for computational treatment, i.e.

$$\mathbf{p}(t) = \mathbf{D}\mathbf{R}^d(t)\,\mathbf{t} \tag{2.4}$$

where $\mathbf{D}$ is a matrix representation of the control points $\mathbf{d}_j$, i.e.

$$\mathbf{D} = \left(\begin{array}{cccc} \mathbf{d}_0 & \mathbf{d}_1 & \cdots & \mathbf{d}_{m-n-2} \end{array} \right).$$

A spline function $R_j^d(t)$ from (2.1) is a polynomial function of degree d that can be represented by its coefficient row vector $\left(\mathbf{r}_j^d(t)\right)^\top$ that is a piecewise constant function of the curve parameter t, and a vector $\mathbf{t}$ of powers of t, i.e.

$$R_j^d = \left(\mathbf{r}_j^d(t)\right)^\top \mathbf{t}, \tag{2.5}$$

where $\mathbf{t} = \left(\begin{array}{cccc} t^n & t^{n-1} & \cdots & t & 1 \end{array} \right)^\top$. Using compact matrix notation in (2.4), the row vectors $\left(\mathbf{r}_j^d\right)^\top$ from (2.5) can be represented as a curve parameter-dependent matrix form, i.e.

$$\mathbf{R}^d(t) = \begin{pmatrix} \left(\mathbf{r}_0^d(t)\right)^\top \\ \left(\mathbf{r}_1^d(t)\right)^\top \\ \vdots \\ \left(\mathbf{r}_{m-n-3}^d(t)\right)^\top \\ \left(\mathbf{r}_{m-n-2}^d(t)\right)^\top \end{pmatrix}.$$

The matrix form of $\mathbf{p}(t)$ from (2.4) on the previous page additionally offers a simple way to compute the derivative of the curve with respect to the curve parameter, denoted as $\mathbf{p}'(t)$ since only the power vector $\mathbf{t}$ needs to be differentiated, i.e.

$$\mathbf{p}'(t) = \mathbf{D}\mathbf{R}^d(t)\,\mathbf{t}'$$

where

$$\mathbf{t}' = \operatorname{diag}(n, n-1, \ldots, 2, 1, 0) \begin{pmatrix} t^{n-1} \\ t^{n-2} \\ \vdots \\ t \\ 1 \\ 0 \end{pmatrix}.$$

The computation of the time-dependent column vector in $\mathbf{t}'$ can be accomplished by shifting up $\mathbf{t}$ and filling the resulting gap with zero.

2.1.1 Properties of NURBS curves

In this section, a selection of properties of NURBS curves that are important for the application in this thesis are presented. The listing below is not exhaustive, a complete version is provided in [10].

- Partition of unity: $\sum_{j=0}^{m-n-2} R_j^d(t) = 1 \quad \forall\, t \in [a, b]$

- Local support: $R_j^d(t) = 0 \quad \text{for} \quad t \notin \left[t_j, t_{j+d+1}\right)$

- Knot multiplicity and continuity: At a knot t_l with multiplicity k_l, $R_j^d\,(t = t_l)$ is $d - k_l$ times continuously differentiable, i.e. $R_j^d\,(t = t_l) \in \mathcal{C}^{d-k_l}$.

- Convex hull: $\mathbf{p}\,(t)$ lies within the control polygon, i.e. the convex hull of the control points $\mathbf{d}_{j-d}, ..., \mathbf{d}_j$ for $t \in [t_j, t_{j+1})$.

  In general, a convex set of two points $\mathbf{x}_i$ and $\mathbf{x}_j$ is their connecting line, i.e.

  $$\mathrm{conv}\,(\mathbf{x}_i, \mathbf{x}_j) = \left\{ \mathbf{x}_i + \lambda\,(\mathbf{x}_j - \mathbf{x}_i) \right\}, \quad 0 \leq \lambda \leq 1.$$

  In the two-dimensional example in Figure 2.1, the set $\mathcal{A}$ is a convex set of all points $\mathbf{x}_i$. In Figure 2.2 this is not the case, $\mathcal{B}$ is only the convex set of points $\mathbf{x}_1$ and $\mathbf{x}_3$, and $\mathbf{x}_2$ and $\mathbf{x}_5$ but not for any other combination of the depicted points. The convex hull of a set of points is the convex set of minimum size. Examples for the convex hull property can be found in Figure 2.5 and Figure 2.6 on page 13.

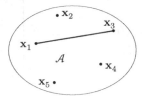

Figure 2.1: Example for a convex set

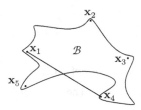

Figure 2.2: Example for a non-convex set

- Local approximation: If a control point $\mathbf{d}_j$ is moved, only the portion of $\mathbf{p}\,(t)$ on the interval $t \in [t_j, t_{j+d+1})$ is affected. An example for the local approximation property can be found in Figure 2.6 on page 13.

2.1.2 Recursion formula

A computation method for NURBS that is particularly interesting for computer implementation is the recursion formula by DE BOOR, COX and MANSFIELD.

The simple case of this recursion is called a base function, which in this case is a constant that is either one or zero, i.e.

$$N_j^0 = \begin{cases} 1 & t_j \leq t < t_{j+1} \\ 0 & \text{otherwise} \end{cases}$$

where $j = 0, \ldots, m-2$. The general case that reduces the functions N_j^d towards the simple case $d = 0$ is

$$N_j^d(t) = \frac{t - t_j}{t_{j+d} - t_j} N_j^{d-1}(t) + \frac{t_{j+d+1} - t}{t_{j+d+1} - t_{j+1}} N_{j+1}^{d-1}(t)$$

where d is the local degree of the polynomial functions, n is the maximum degree of $N_j^d(t)$ and m is the number of knots. For knot multiplicities greater than zero, fractions $\frac{0}{0}$ may occur, which are defined to be zero, i.e. $\frac{0}{0} := 0$.

Following (2.3) on page 6, the control points $\mathbf{d}_j$ can be weighted using factors w_j. From the unity property from Section 2.1.1 follows that $R_j^d = N_j^d$ if all weights are equal, i.e. $w_i = w_j, \forall i, j$. If all weighting factors are one, the curve is called a B-spline curve.

Example

In this example, the quadratic ($n = 2$) basis functions N_j^d and the rational functions R_j^d are to be computed and the NURBS curve $\mathbf{p}(t)$ is to be visualized for the knots in the knot vector $\mathbf{T}$. This example is based on Exercise 4.1 from [10].

$$\mathbf{T} = \begin{pmatrix} t_0 & t_1 & \ldots & t_6 & t_7 \end{pmatrix}$$
$$= \begin{pmatrix} 0 & 0 & 0 & \frac{1}{3} & \frac{2}{3} & 1 & 1 & 1 \end{pmatrix}$$

$$\mathbf{d}_0 = \begin{pmatrix} 0 \\ 0 \end{pmatrix} \quad w_0 = 1$$

$$\mathbf{d}_1 = \begin{pmatrix} 1 \\ 1 \end{pmatrix} \quad w_1 = 4 \qquad \mathbf{d}_3 = \begin{pmatrix} 4 \\ 1 \end{pmatrix} \quad w_3 = 1$$

$$\mathbf{d}_2 = \begin{pmatrix} 3 \\ 2 \end{pmatrix} \quad w_2 = 1 \qquad \mathbf{d}_4 = \begin{pmatrix} 5 \\ -1 \end{pmatrix} \quad w_4 = 1$$

Following the DE BOOR, COX and MANSFIELD algorithm presented in Section 2.1.2, the basis functions are found to be a scheme as depicted in Figure 2.3 on the next page where $\forall\, j \in \{0, 1, 2, 3, 4\}$

$$N_j^0(t) = \begin{cases} 1 & t_j \leq t < t_{j+1} \\ 0 & \text{otherwise} \end{cases} \tag{2.6}$$

$$N_{j+1}^0(t) = \begin{cases} 1 & t_{j+1} \leq t < t_{j+2} \\ 0 & \text{otherwise} \end{cases} \tag{2.7}$$

$$N_{j+2}^0(t) = \begin{cases} 1 & t_{j+2} \leq t < t_{j+3} \\ 0 & \text{otherwise} \end{cases} \tag{2.8}$$

$$N_j^1(t) = \frac{t - t_j}{t_{j+1} - t_j} N_j^0(t) + \frac{t_{j+2} - t}{t_{j+2} - t_{j+1}} N_{j+1}^0(t) \tag{2.9}$$

$$N_{j+1}^1(t) = \frac{t - t_{j+1}}{t_{j+2} - t_{j+1}} N_{j+1}^0(t) + \frac{t_{j+3} - t}{t_{j+3} - t_{j+2}} N_{j+2}^0(t) \tag{2.10}$$

$$N_j^2(t) = \frac{t - t_j}{t_{j+2} - t_j} N_j^1(t) + \frac{t_{j+3} - t}{t_{j+3} - t_{j+1}} N_{j+1}^1(t). \tag{2.11}$$

Substituting (2.6) to (2.8) on the previous page in (2.9) and (2.10) yields

$$N_j^1(t) = \begin{cases} \frac{t-t_j}{t_{j+1}-t_j} & t_j \leq t < t_{j+1} \\ \frac{t_{j+2}-t}{t_{j+2}-t_{j+1}} & t_{j+1} \leq t < t_{j+2} \\ 0 & \text{otherwise} \end{cases} \tag{2.12}$$

$$N_{j+1}^1(t) = \begin{cases} \frac{t-t_{j+1}}{t_{j+2}-t_{j+1}} & t_{j+1} \leq t < t_{j+2} \\ \frac{t_{j+3}-t}{t_{j+3}-t_{j+2}} & t_{j+2} \leq t < t_{j+3} \\ 0 & \text{otherwise} \end{cases} \cdot \tag{2.13}$$

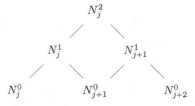

Figure 2.3: NURBS example — Scheme of B-spline basis functions N_j^d

Substituting (2.12) and (2.13) in (2.11) yields

$$N_j^2(t) = \begin{cases} \frac{t-t_j}{t_{j+2}-t_j}\frac{t-t_j}{t_{j+1}-t_j} & t_j \leq t < t_{j+1} \\ \frac{t-t_j}{t_{j+2}-t_j}\frac{t_{j+2}-t}{t_{j+2}-t_{j+1}} + \frac{t_{j+3}-t}{t_{j+3}-t_{j+1}}\frac{t-t_{j+1}}{t_{j+2}-t_{j+1}} & t_{j+1} \leq t < t_{j+2} \\ \frac{t_{j+3}-t}{t_{j+3}-t_{j+1}}\frac{t_{j+3}-t}{t_{j+3}-t_{j+2}} & t_{j+2} \leq t < t_{j+3} \\ 0 & \text{otherwise} \end{cases} \quad \forall\, j = 0\ldots 4.$$

Figure 2.4 on the next page shows the B-spline basis functions N_j^d for $j = 0\ldots m-n-2 = 4$ and $d = 0\ldots n = 2$.

The weighted functions R_j^d can be easily computed using (2.3) on page 6. Following (2.1), the NURBS function $\mathbf{p}(t)$ can now be obtained and visualized, see Figure 2.5.

To show the local approximation property from Section 2.1.1, one of the control points

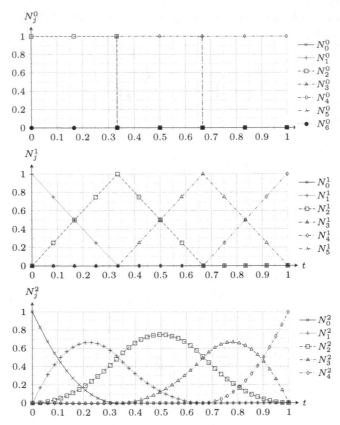

Figure 2.4: NURBS example — B-spline basis function N_j^d

from the example is modified, i.e.

$$\mathbf{d}_3 = \begin{pmatrix} 4.5 \\ 1.5 \end{pmatrix}.$$

The result is depicted in Figure 2.6 on the next page where the modified curve $\mathbf{p}_{\mathrm{mod}}(t)$ only differs from the original curve $\mathbf{p}(t)$ in the area between the adjoining control points $\mathbf{d}_2$ and $\mathbf{d}_4$.

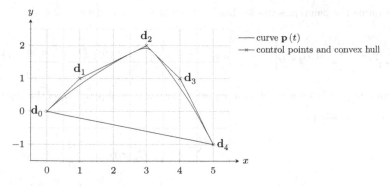

Figure 2.5: NURBS example — curve

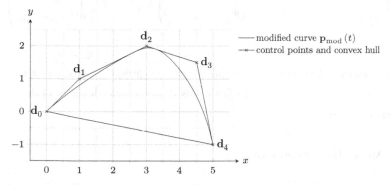

Figure 2.6: NURBS example — modified curve

2.1.3 Curve approximation

For the approximation of a series of $N \geq m - n - 1$ points $\mathbf{p}_i$ at parameters $\tau_i \in [a, b]$ with weights w_k where $k \in \{0, 1, ..., m - n - 3, m - n - 2\}$ a NURBS curve $\tilde{\mathbf{p}}(t)$ of degree d with m knots t_k where $k \in \{0, 1, ..., m - 2, m - 1\}$ is to be found such that $\tilde{\mathbf{p}}(t = \tau_i) \approx \mathbf{p}_i$, see [7].

The goal is to minimize the sum of error squares of the approximation $\tilde{\mathbf{p}}$ with respect to the given points $\mathbf{p}_i$ at the given parameters τ_i, i.e.

$$\sum_{i=0}^{N-1} |\mathbf{p}_i - \tilde{\mathbf{p}}(t = \tau_i)|^2 \quad \rightarrow \quad \min, \quad \tau_i \in [a, b], \quad i \in \{0, 1, ..., N - 2, N - 1\}.$$

Substituting the sum representation from (2.1) on page 5 for $\tilde{\mathbf{p}}$ yields

$$\sum_{i=0}^{N-1} \left| \mathbf{p}_i - \sum_{j=0}^{m-n-2} \mathbf{d}_j R_j^d (\tau_i) \right|^2 \quad \rightarrow \quad \min, \quad \tau_i \in [a,b], \quad i \in \{0, 1, \dots, N-2, N-1\}.$$

Minimization yields a system of linear equations for the control points $\mathbf{d}_j$ of the approximated curve points $\tilde{\mathbf{p}}$,

$$\sum_{j=0}^{m-n-2} \sum_{i=0}^{N-1} R_l^d (\tau_i) R_j^d (\tau_i) \mathbf{d}_j = \sum_{i=0}^{N-1} R_l^d (\tau_i) \mathbf{p}_i$$

$$\begin{pmatrix} \sum\limits_{i=0}^{N-1} R_0^d R_0^d & \sum\limits_{i=0}^{N-1} R_0^d R_1^d & \cdots & \sum\limits_{i=0}^{N-1} R_0^d R_{m-n-2}^d \\ \sum\limits_{i=0}^{N-1} R_1^d R_j^d & \sum\limits_{i=0}^{N-1} R_1^d R_j^d & \cdots & \sum\limits_{i=0}^{N-1} R_1^d R_{m-n-2}^d \\ \vdots & \vdots & \ddots & \vdots \\ \sum\limits_{i=0}^{N-1} R_{m-n-2}^d R_0^d & \sum\limits_{i=0}^{N-1} R_{m-n-2}^d R_1^d & \cdots & \sum\limits_{i=0}^{N-1} R_{m-n-2}^d R_{m-n-2}^d \end{pmatrix} \begin{pmatrix} \mathbf{d}_0 \\ \mathbf{d}_1 \\ \vdots \\ \mathbf{d}_{m-n-2} \end{pmatrix} = \begin{pmatrix} \sum\limits_{i=0}^{N-1} R_0^d \mathbf{p}_i \\ \sum\limits_{i=0}^{N-1} R_1^d \mathbf{p}_i \\ \vdots \\ \sum\limits_{i=0}^{N-1} R_{m-n-2}^d \mathbf{p}_i \end{pmatrix}$$

where $l \in \{0, 1, \dots, m-n-2\}$ and the argument τ_i of R was suppressed for the sake of compactness. The resulting band matrix can be efficiently solved for each of the components of $\mathbf{d}_j$. If any of the control points $\mathbf{d}_j$ are known, the order of the system of equation is reduced.

2.2   Application in optimization

A NURBS curve is defined by its maximum degree n, its range $t \in [a,b]$ with m knots t_k, its control points $\mathbf{d}_j$ and its weights w_j. The maximum degree and the range are mostly determined by the requirements of the application itself such as the continuity level. The number of knots and the knot positions, control points and the weights are possible parameters for optimization. The number and positions of knots can be adjusted in order to allow better local adaptivity while the control points and their weights influence the general shape of the curve. Due to the local approximation property from Section 2.1.1, local shape adjustments are possible. For some applications, such as the use as a time-dependent polynomial in the equations of motion of a manipulator, derivatives of the polynomial are necessary to compute. The matrix representation of a NURBS curve derived in (2.4) on page 6 allows to obtain those derivatives in a computationally cheap manner.

3. Modeling: Kinematics and Dynamics of Redundant Robots

In this chapter the *Projection Equation*, a synthetic method for deriving the equations of motion of a multibody system, will be briefly described and applied to an example. A comprehensive discussion of this method is presented in [2].

3.1 Projection Equation

A subsystem formulation of the *Projection Equation* is denoted as

$$
\sum_{n=1}^{N_{\text{sub}}} \underbrace{\left(\frac{\partial \dot{\mathbf{y}}_n}{\partial \dot{\mathbf{s}}}\right)^{\mathsf{T}}}_{\mathbf{F}_n^{\mathsf{T}}} \sum_{i=1}^{N_n} \left(\underbrace{\mathrm{R}\left(\left(\frac{\partial \mathbf{v}_c}{\partial \dot{\mathbf{y}}_n}\right)^{\mathsf{T}} \quad \left(\frac{\partial \omega_c}{\partial \dot{\mathbf{y}}_n}\right)^{\mathsf{T}} \right)}_{\mathbf{F}_i^{\mathsf{T}}} {}_{\mathrm{R}}\left(\begin{array}{c} \dot{\mathbf{p}} + \tilde{\omega}_{\mathrm{IR}}\mathbf{p} - \mathbf{f}^{\mathrm{e}} \\ \dot{\mathbf{L}} + \tilde{\omega}_{\mathrm{IR}}\mathbf{L} - \mathbf{M}^{\mathrm{e}} \end{array} \right) \right)_i = 0 \qquad (3.1)
$$

with subscript R suppressed. This matrix equation describes a structure of N_{sub} subsystems wherein the n-th subsystem consists of N_n bodies. In (3.1) $\dot{\mathbf{s}}$ describes the system's generalized velocities and $\dot{\mathbf{y}}_n$ the auxiliary velocities of the n-th subsystem. $\mathbf{p}$ is the linear momentum, $\mathbf{L}$ the angular momentum and $\mathbf{f}^{\mathrm{e}}$ and $\mathbf{M}^{\mathrm{e}}$ are the external forces and moments applied to body i of the n-th subsystem, respectively. ω_{IR} denotes the angular velocity of the coordinate frame fixed to body i with respect to the inertial frame. This shows that for each body (or subsystem) a separate coordinate frame can be chosen. It

15

is possible to express the linear momentum conservation law in a coordinate system that is different from the system the angular momentum conservation law is described in. Another advantage of the *Projection Equation* over conventional analytic methods is that non-holonomic velocities can be incorporated.

As shown in [2], close investigation of a subsystem expression yields the velocities of the center of mass for the i-th body in the n-th subsystem

$$\dot{\mathbf{y}} = \begin{pmatrix} \mathbf{v}_c \\ \omega_c \end{pmatrix}_i = \begin{pmatrix} \mathbf{v}_0 + \tilde{\omega}_0 \mathbf{r}_c + \dot{\mathbf{r}}_c \\ \omega_0 + \omega_r \end{pmatrix}_i.$$

In

$$\dot{\mathbf{y}}_n = \begin{pmatrix} \mathbf{v}_0^\top & \omega_0^\top & \dot{\mathbf{r}}_c^\top & \omega_r^\top \end{pmatrix}^\top$$

$\mathbf{v}_0$ and ω_0 denote the translational and angular velocities of the origin of the coordinate frame fixed to the body. The translational velocity $\dot{\mathbf{r}}_c$ and the angular velocity ω_r describe the relative motion of the center of mass of body i with respect to the body coordinate frame. $\dot{\mathbf{y}}_n$ denote the auxiliary velocities of the n-th subsystem, the functional matrix $\overline{\mathbf{F}}_i$ is found to be

$$\overline{\mathbf{F}}_i^\top = \begin{pmatrix} \left(\frac{\partial \mathbf{v}_c}{\partial \dot{\mathbf{y}}_n}\right)^\top & \left(\frac{\partial \omega_c}{\partial \dot{\mathbf{y}}_n}\right)^\top \end{pmatrix}_i = \begin{pmatrix} \mathbf{I} & \tilde{\mathbf{r}}_c^\top & \mathbf{I} & \mathbf{O} \\ \mathbf{O} & \mathbf{I} & \mathbf{O} & \mathbf{I} \end{pmatrix}_i^\top,$$

considering that $\tilde{\omega}_0 \mathbf{r}_c = \tilde{\mathbf{r}}_c^\top \omega_0$. With the i-th body's linear and angular momentum

$$\begin{pmatrix} \mathbf{p} \\ \mathbf{L} \end{pmatrix}_i = \begin{pmatrix} m\mathbf{I} & \mathbf{O} \\ \mathbf{O} & \mathbf{J}_c \end{pmatrix}_i \begin{pmatrix} \mathbf{v}_c \\ \omega_c \end{pmatrix}_i = \overline{\mathbf{M}}_i \overline{\mathbf{F}}_i \dot{\mathbf{y}}_n,$$

wherein m denotes the body's mass and $\mathbf{J}_c$ is the inertia tensor of the body with respect to its center of mass, and the respective time derivatives

$$\begin{pmatrix} \dot{\mathbf{p}} \\ \dot{\mathbf{L}} \end{pmatrix}_i = \dot{\overline{\mathbf{M}}}_i \overline{\mathbf{F}}_i \dot{\mathbf{y}}_n + \overline{\mathbf{M}}_i \dot{\overline{\mathbf{F}}}_i \dot{\mathbf{y}}_n + \overline{\mathbf{M}}_i \overline{\mathbf{F}}_i \ddot{\mathbf{y}}_n$$

the equations of motion for the i-th body in the n-th subsystem expressed in the auxiliary velocities $\dot{\mathbf{y}}_n$ can be completed, i.e.

$$
\left(\begin{pmatrix} \left(\frac{\partial \mathbf{v}_c}{\partial \dot{\mathbf{y}}_n} \right)^{\top} & \left(\frac{\partial \omega_c}{\partial \dot{\mathbf{y}}_n} \right)^{\top} \end{pmatrix} \begin{pmatrix} \dot{\mathbf{p}} + \tilde{\omega}_{\mathrm{IR}} \mathbf{p} - \mathbf{f}^{\mathrm{e}} \\ \dot{\mathbf{L}} + \tilde{\omega}_{\mathrm{IR}} \mathbf{L} - \mathbf{M}^{\mathrm{e}} \end{pmatrix} \right)_i
$$
$$
= \overline{\mathbf{F}}_i^{\top} \left(\overline{\mathbf{M}}_i \overline{\mathbf{F}}_i \dot{\mathbf{y}}_n + \overline{\dot{\mathbf{M}}}_i \overline{\mathbf{F}}_i \dot{\mathbf{y}}_n + \overline{\mathbf{M}}_i \overline{\dot{\mathbf{F}}}_i \dot{\mathbf{y}}_n + \right.
$$
$$
\left. + \mathrm{blockdiag}\,(\tilde{\omega}_{\mathrm{IR}}, \tilde{\omega}_{\mathrm{IR}})_i\, \overline{\mathbf{M}}_i \overline{\mathbf{F}}_i \dot{\mathbf{y}}_n - \begin{pmatrix} \mathbf{f}^{\mathrm{e}} & \mathbf{M}^{\mathrm{e}} \end{pmatrix}^{\top}_i \right).
$$

Rewriting the equation yields a well-known matrix form of the equations of motion,

$$
= \overline{\mathbf{F}}_i^{\top} \overline{\mathbf{M}}_i \overline{\mathbf{F}}_i \ddot{\mathbf{y}}_n + \overline{\mathbf{F}}_i^{\top} \left(\overline{\dot{\mathbf{M}}}_i \overline{\mathbf{F}}_i + \overline{\mathbf{M}}_i \overline{\dot{\mathbf{F}}}_i + \mathrm{blockdiag}\,(\tilde{\omega}_{\mathrm{IR}}, \tilde{\omega}_{\mathrm{IR}})_i\, \overline{\mathbf{M}}_i \overline{\mathbf{F}}_i \right) \dot{\mathbf{y}}_n - \mathbf{Q}_i^{\mathrm{e}}
$$
$$
= \mathbf{M}_i \ddot{\mathbf{y}}_n + \mathbf{G}_i \dot{\mathbf{y}}_n - \mathbf{Q}_i^{\mathrm{e}},
$$

where $\mathbf{Q}_i^{\mathrm{e}}$ is the vector of external forces and moments applied to the body, i.e.

$$
\mathbf{Q}_i^{\mathrm{e}} = \overline{\mathbf{F}}_i^{\top} \begin{pmatrix} \mathbf{f}^{\mathrm{e}} \\ \mathbf{M}^{\mathrm{e}} \end{pmatrix}_i .
$$

Finally, the subsystems can be assembled according to (3.1) on page 15, rewritten as

$$
\sum_{n=1}^{N_{\mathrm{sub}}} \mathbf{F}_n^{\top} \sum_{i=1}^{N_n} \left(\mathbf{M}_i \ddot{\mathbf{y}}_n + \mathbf{G}_i \dot{\mathbf{y}}_n - \mathbf{Q}_i^{\mathrm{e}} \right) = \mathbf{0} \tag{3.2}
$$

and in minimal description

$$
\mathbf{M} \ddot{\mathbf{s}} + \mathbf{G} - \mathbf{Q}^{\mathrm{e}} = \mathbf{0}
$$

where

$$
\mathbf{M} = \sum_{n=1}^{N_{\mathrm{sub}}} \mathbf{F}_n^{\top} \sum_{i=1}^{N_n} \mathbf{M}_i \mathbf{F}_n
$$
$$
\mathbf{G} = \sum_{n=1}^{N_{\mathrm{sub}}} \mathbf{F}_n^{\top} \sum_{i=1}^{N_n} \mathbf{G}_i \mathbf{F}_n
$$
$$
\mathbf{Q}^{\mathrm{e}} = \sum_{n=1}^{N_{\mathrm{sub}}} \mathbf{F}_n^{\top} \sum_{i=1}^{N_n} \mathbf{Q}_i^{\mathrm{e}} .
$$

Considering the kinematic relationship between one subsystem and its predecessing subsystem,

$$\dot{\mathbf{y}}_n = \mathbf{T}_{n,n-1}\dot{\mathbf{y}}_{n-1} + \mathbf{F}_n\dot{\mathbf{s}}_n, \tag{3.3}$$

referred to as *kinematic chain* in [2], each subsystem's auxiliary velocities $\dot{\mathbf{y}}_n$ can be alternatively expressed using the generalized velocities $\dot{\mathbf{s}}$ and the predecessor's auxiliary velocities $\dot{\mathbf{y}}_{n-1}$,

$$\begin{pmatrix} {}_n\mathbf{v}_{0,n} \\ {}_n\omega_{0,n} \\ {}_n\dot{\mathbf{r}}_n \\ {}_n\omega_n \end{pmatrix} = \underbrace{\left(\begin{pmatrix} \mathbf{A}_{n,n-1} & \mathbf{O} \\ \mathbf{O} & \mathbf{A}_{n,n-1} \end{pmatrix} \cdot \begin{pmatrix} \mathbf{I} & {}_{n-1}\tilde{\mathbf{r}}_{n-1}^{\top} & \mathbf{I} & {}_{n-1}\tilde{\mathbf{r}}_{n-1}^{\top} \\ \mathbf{O} & \mathbf{I} & \mathbf{O} & \mathbf{I} \end{pmatrix} \right)}_{\mathbf{T}_{n,n-1}} \underbrace{\begin{pmatrix} {}_{n-1}\mathbf{v}_{0,n-1} \\ {}_{n-1}\omega_{0,n-1} \\ {}_{n-1}\dot{\mathbf{r}}_{n-1} \\ {}_{n-1}\omega_{n-1} \end{pmatrix}}_{\dot{\mathbf{y}}_{n-1}} + \mathbf{F}_n\dot{\mathbf{s}}_n.$$

3.2 Modeling example: subsystem motor, gear, arm

In order to clarify the method described above, the synthetic modeling approach will be utilized in the following example. The subsystem consisting of motor, gear and arm

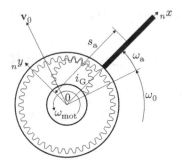

Figure 3.1: Subsystem motor, gear, arm

as depicted in Figure 3.1 is considered mounted on a predecessor body or subsystem at point 0. A frame of reference (subscript n) is defined to describe the subsystem's motion. This frame of reference moves with the translational velocity $\mathbf{v}_0$ and with the rotational velocity ω_0 with respect to the inertial origin. For the subsystem the auxiliary velocities are defined as

$$\dot{\mathbf{y}}_n = \begin{pmatrix} \mathbf{v}_0 \\ \omega_0 \\ \omega_a \end{pmatrix}.$$

The relative translational velocity $\dot{\mathbf{r}}_c$ and the first two components of the relative rotational velocity ω_r are zero and therefore omitted for the sake of compactness.

The subsystem's frame of reference moves with its absolute translational velocity

$$\mathbf{v}_0 = \begin{pmatrix} v_{0,x} & v_{0,y} & v_{0,z} \end{pmatrix}^{\mathsf{T}}$$

and its absolute rotational velocity

$$\omega_0 = \begin{pmatrix} \omega_{0,x} & \omega_{0,y} & \omega_{0,z} \end{pmatrix}^{\mathsf{T}}.$$

Body arm

The body's local coordinate system is a principal coordinate system. Its local motion can be described by the rotational and translational velocity of its center of mass,

$$\omega_{\mathrm{arm}} = \omega_0 + \begin{pmatrix} 0 \\ 0 \\ \omega_a \end{pmatrix}, \quad \mathbf{v}_{\mathrm{arm}} = \mathbf{v}_0 + \tilde{\omega}_{\mathrm{arm}} \begin{pmatrix} s_a \\ 0 \\ 0 \end{pmatrix},$$

respectively. With the body's velocities

$$\dot{\mathbf{y}}_{\mathrm{arm}} = \begin{pmatrix} \mathbf{v}_{\mathrm{arm}} \\ \omega_{\mathrm{arm}} \end{pmatrix},$$

the functional matrix

$$\overline{\mathbf{F}}_{\text{arm}} = \frac{\partial \dot{\overline{\mathbf{y}}}_{\text{arm}}}{\partial \dot{\mathbf{y}}_n} = \begin{pmatrix} 1 & 0 & 0 & 0 & 0 & 0 & 0 \\ 0 & 1 & 0 & 0 & 0 & s_{\text{a}} & s_{\text{a}} \\ 0 & 0 & 1 & 0 & -s_{\text{a}} & 0 & 0 \\ 0 & 0 & 0 & 1 & 0 & 0 & 0 \\ 0 & 0 & 0 & 0 & 1 & 0 & 0 \\ 0 & 0 & 0 & 0 & 0 & 1 & 1 \end{pmatrix},$$

that projects from body space to subsystem space, can be derived. Now the body's local mass matrix, which is a constant diagonal matrix,

$$\overline{\mathbf{M}}_{\text{arm}} = \text{diag}\left(m_{\text{arm}}, m_{\text{arm}}, m_{\text{arm}}, A_{\text{c,arm}}, B_{\text{c,arm}}, C_{\text{c,arm}}\right)$$

and the body's gyroscopic matrix

$$\mathbf{G}_{\text{arm}} = \overline{\mathbf{F}}_{\text{arm}}^{\top}\left(\underbrace{\dot{\overline{\mathbf{M}}}_{\text{arm}}\overline{\mathbf{F}}_{\text{arm}} + \overline{\mathbf{M}}_{\text{arm}}\dot{\overline{\mathbf{F}}}_{\text{arm}}}_{\mathbf{O}} + \text{blockdiag}\left(\tilde{\omega}_0, \tilde{\omega}_0\right)_{\text{arm}}\overline{\mathbf{M}}_{\text{arm}}\overline{\mathbf{F}}_{\text{arm}}\right)$$

can be derived wherein the first two terms vanish as $\dot{\overline{\mathbf{M}}}_{\text{arm}} = \mathbf{O}$ and $\dot{\overline{\mathbf{F}}}_{\text{arm}} = \mathbf{O}$. The vector of forces and moments applied to the arm is

$$\mathbf{Q}_{\text{arm}}^{\text{e}} = \begin{pmatrix} 0 & 0 & 0 & 0 & 0 & 0 & -d\omega_{\text{a}} \end{pmatrix}^{\top},$$

wherein d refers to the coefficient of velocity-based kinetic friction in the arm joint.

Body motor

This body also translates with velocity $\mathbf{v}_{\text{mot}} = \mathbf{v}_0$ and rotates with the angular velocity

$$\omega_{\text{mot}} = \omega_0 + \begin{pmatrix} 0 \\ 0 \\ \omega_{\text{a}}{}^{i}_{\text{G}} \end{pmatrix},$$

where ω_a is the arm's relative angular velocity multiplied with the constant gear ratio i_G. The body's auxiliary velocities are set to be

$$\dot{\overline{y}}_{mot} = \begin{pmatrix} v_{mot} \\ \omega_{mot} \end{pmatrix}.$$

With the functional matrix

$$\overline{F}_{mot} = \frac{\partial \dot{\overline{y}}_{mot}}{\partial \dot{y}_n} = \begin{pmatrix} 1 & 0 & 0 & 0 & 0 & 0 & 0 \\ 0 & 1 & 0 & 0 & 0 & 0 & 0 \\ 0 & 0 & 1 & 0 & 0 & 0 & 0 \\ 0 & 0 & 0 & 1 & 0 & 0 & 0 \\ 0 & 0 & 0 & 0 & 1 & 0 & 0 \\ 0 & 0 & 0 & 0 & 0 & 1 & i_G \end{pmatrix}$$

and the body's mass matrix

$$\overline{M}_{mot} = \text{diag}\left(m_{mot}, m_{mot}, m_{mot}, A_{c,mot}, B_{c,mot}, C_{c,mot}\right),$$

the gyroscopic matrix

$$G_{mot} = \overline{F}_{mot}^{\top} \left(\underbrace{\dot{\overline{M}}_{mot} \overline{F}_{mot} + \overline{M}_{mot} \dot{\overline{F}}_{mot}}_{O} + \text{blockdiag}\left(\tilde{\omega}_0, \tilde{\omega}_0\right)_{mot} \overline{M}_{mot} \overline{F}_{mot} \right)$$

can be derived. The vector of forces and moments applied to the motor is

$$Q_{mot}^e = \begin{pmatrix} 0 & 0 & 0 & 0 & 0 & 0 & i_G M_{mot} \end{pmatrix}^{\top}$$

wherein M_{mot} denotes the motor torque.

System assemblage

Now that the functional matrices $\overline{F}_i$, the mass matrices $\overline{M}_i$ and the gyroscopic matrices $\overline{G}_i$ are available for both bodies $i = \{\text{arm}, \text{mot}\}$, they can be assembled to form the subsystem's respective matrices,

$$\mathbf{M}_n = \overline{\mathbf{F}}_{\text{arm}}^{\top}\overline{\mathbf{M}}_{\text{arm}}\overline{\mathbf{F}}_{\text{arm}} + \overline{\mathbf{F}}_{\text{mot}}^{\top}\overline{\mathbf{M}}_{\text{mot}}\overline{\mathbf{F}}_{\text{mot}}$$

$$\mathbf{G}_n = \mathbf{G}_{\text{arm}} + \mathbf{G}_{\text{mot}}.$$

The vector of forces and moments acting on the subsystem is

$$\mathbf{Q}_n^{\text{e}} = \mathbf{Q}_{\text{arm}}^{\text{e}} + \mathbf{Q}_{\text{mot}}^{\text{e}}.$$

In [2] an example is presented where the *Projection Equation* is used to derive the equations of motion for a similar subsystem where elasticity is added to the gears.

4. Approaches to Minimum-Time Trajectory Planning

This section describes the problem of minimum-time trajectory planning along predefined paths for serial robots, especially for redundant manipulators. Inverse kinematics are used in order to utilize null space motion and thus exploit a robot's kinematic redundancy for a further reduction of the trajectory end time.

For a robot's end-effector following a known path $\mathbf{r}_{E,d}$ and a specified end-effector orientation $\varphi_{E,d}$, i.e.

$$\mathbf{z}_{E,d} = \begin{pmatrix} \mathbf{r}_{E,d} \\ \varphi_{E,d} \end{pmatrix},$$

joint trajectories $\mathbf{q}\left(\mathbf{z}_{E,d}\left(\mathbf{x}\right)\right)$, expressed by a set of parameters denoted in vector form as $\mathbf{x}$, are to be determined. Additional requirements to such trajectories w.r.t. the system can be imposed by the optimization problem below.

4.1 Minimum-time optimization problem

In general, optimization problems regarding minimum-time trajectories lead to the search for optimal solutions in infinite dimensional function space. In this case, the problem is reduced such that the trajectory is described using a finite set of trajectory parameters denoted as $\mathbf{x}$. A set of optimal parameters $\mathbf{x}^*$ that minimizes the trajectory end time t_E is obtained with the optimization problem

$$\mathbf{x}^* = \arg \min_{\mathbf{x}} f(\mathbf{x}) \tag{4.1}$$

$$\mathrm{s.t.} \quad \mathbf{c}_\mathrm{eq} = \mathbf{0}$$

$$\mathbf{c}_\mathrm{ineq} \leq \mathbf{0}$$

with the cost function

$$f(\mathbf{x}) = t_\mathrm{E}(\mathbf{x}).$$

The optimization problem above is subject to the dynamic system and a path-following equality constraint

$$\mathbf{c}_\mathrm{eq} = \mathbf{z}_\mathrm{E,d} - \mathbf{z}_\mathrm{E} = \mathbf{0}$$

or the end-effector path $\mathbf{z}_\mathrm{E,d}$ is described as

$$\mathbf{z}_\mathrm{E,d} = \mathbf{z}_\mathrm{E,d}(s) \quad s \in [0,1], \quad s(t=0) = 0, \quad s(t=t_\mathrm{E}) = 1$$

in combination with the equality constraint. Inequality constraints such as e.g. constrained joint velocities $\dot{\mathbf{q}}\left(\mathbf{z}_\mathrm{E,d}(s), \dot{\mathbf{z}}_\mathrm{E,d}(s,\dot{s})\right)$ or accelerations $\ddot{\mathbf{q}}\left(\mathbf{z}_\mathrm{E,d}(s), \dot{\mathbf{z}}_\mathrm{E,d}(s,\dot{s}), \ddot{\mathbf{z}}_\mathrm{E,d}(s,\dot{s},\ddot{s})\right)$, i.e.

$$\mathbf{c}_{\mathrm{ineq},\dot{\mathbf{q}}}(\mathbf{x}) = \begin{pmatrix} \dot{\mathbf{q}} - \dot{\mathbf{q}}_\mathrm{max} \\ \dot{\mathbf{q}}_\mathrm{min} - \dot{\mathbf{q}} \end{pmatrix} \leq \mathbf{0}, \qquad \mathbf{c}_{\mathrm{ineq},\ddot{\mathbf{q}}}(\mathbf{x}) = \begin{pmatrix} \ddot{\mathbf{q}} - \ddot{\mathbf{q}}_\mathrm{max} \\ \ddot{\mathbf{q}}_\mathrm{min} - \ddot{\mathbf{q}} \end{pmatrix} \leq \mathbf{0},$$

or constrained input actions $\mathbf{Q}$, i.e.

$$\mathbf{c}_{\mathrm{ineq},\mathbf{Q}}(\mathbf{x}) = \begin{pmatrix} \mathbf{Q} - \mathbf{Q}_\mathrm{max} \\ \mathbf{Q}_\mathrm{min} - \mathbf{Q} \end{pmatrix} \leq \mathbf{0},$$

where the argument $\mathbf{x}$ was omitted in the subvectors, can also be imposed on the optimization problem. The input actions $\mathbf{Q} = \mathbf{Q}(s, \dot{s}, \ddot{s})$ are computed by means of inverse dynamics. The resulting trajectory is required to satisfy those constraints throughout the time interval of the trajectory $[0, t_\mathrm{E}]$.

Similarly, additional restrictions on the path-tracking velocity $\dot{r}_E$ or acceleration $\ddot{r}_E$ can be imposed by means of similar inequality constraints.

In the literature, a well-known approach to the task of finding minimum-time trajectories is described with the cost function

$$f(\mathbf{x}) = \int_0^{t_E} dt = \int_0^1 \frac{1}{\dot{s}} ds$$

with the substitution

$$\dot{s} = \frac{ds}{dt} \quad \rightarrow \quad dt = \frac{ds}{\dot{s}},$$

resulting in an optimal control problem formulation. However, for trajectories where $\exists t_x \in [0, t_E]$ no motion along the path is performed, i.e. $\dot{s}(t = t_x) = 0$, this formulation yields an infinite cost function. This problem can be prevented by measures such as avoiding $\dot{s} = 0$ in the path planning task, excluding any t_x from the cost function, or approximating $\dot{s} = 0$ with very small values of $\dot{s}$ such that possible friction of the system effectively cancels the motion along the path. In contrast to the application of the cost function formulation above, the separation methods in this chapter are applicable to arbitrary trajectories including the case of $\dot{s} = 0$.

4.2 Optimization basics

In this section, a very short overview of a possible strategy to solve the above problem will be given. Detailed information can be obtained from optimization literature such as [6].

Since the above constraints are in general non-linear with respect to the optimization parameters $\mathbf{x}$, the optimization problem is categorized as *inequality-constrained non-linear optimization*, [6].

A method to solve problems of this class, the *Sequential Quadratic Programming (SQP)* method is suggested in [6] because of its efficiency. In each iteration a second-order approximation of the LAGRANGE function is derived in order to obtain a solution from a *Quadratic Programming* subproblem, see [14].

For handling inequality constraints, the used SQP method is based on an *Active Set* algorithm that only incorporates currently active constraints.

4.3 Algorithms for redundant manipulators

If a robot features more joint degrees of freedom than needed to assume a certain task space position it is considered kinematically redundant. In the following section, inverse kinematics solution algorithms will be derived using optimization-based methods. The presented formulations are directly applicable to redundant serial robots and the solutions will be enhanced in order to utilize redundancy to improve particular properties of the instantaneous robot pose such as kinematic and dynamic manipulability. Furthermore, a manipulability measure that exploits the knowledge about the given geometric path of the manipulator's end-effector will be introduced. Finally, a B-spline-based approach to finding minimum-time trajectories will be presented. Its practicability will be investigated by means of the example of a planar three-link manipulator in Section 6.1.

4.3.1 First-order inverse kinematics algorithms

In this section two first-order (velocity-based) methods for computing inverse kinematics solutions for redundant manipulators will be presented.

Minimum-velocity approach

The following approach from [5] solves an optimization problem subject to direct kinematics and a cost function that minimizes velocities.

The cost function to be minimized is chosen to be

$$Z(\dot{\mathbf{q}}) = \frac{1}{2}\dot{\mathbf{q}}^\top \mathbf{W}\dot{\mathbf{q}}$$

wherein $\mathbf{W} = \operatorname{diag}(w_1, \dots, w_n) > 0$ penalizes the components of the vector of joint velocities $\dot{\mathbf{q}} \in \mathbb{R}^n$. Components of $\mathbf{W}$ that are greater than others result in smaller values of the corresponding entries of $\dot{\mathbf{q}}$ than those that are treated with smaller weights.

The minimization of the cost function $Z(\dot{q})$ is equally constrained by the path-tracking requirement of the desired end-effector path $\dot{r}_{E,d} \in \mathbb{R}^m$,

$$\dot{r}_{E,d} - J(q)\,\dot{q} = 0, \tag{4.2}$$

where $J(q)$ denotes the Jacobian matrix, i.e. the projector from joint space velocities to task space velocities. (4.2) is a simplified form of the general path-tracking constraint $\dot{z}_{E,d} - J(q)\,\dot{q} = 0$ that also incorporates the end-effector rotation. The general formulation, however, limits the application of techniques to exploit a manipulator's kinematic redundancy using kinematic manipulabilty, see Section 4.3.4.

The LAGRANGE function L of the optimization problem is found to be

$$L(\dot{q}) = Z(\dot{q}) + \lambda^\top \left(J(q)\,\dot{q} - \dot{r}_{E,d} \right)$$

wherein λ denotes the vector of LAGRANGE multipliers. Rewriting the KKT (KARUSH-KUHN-TUCKER) conditions

$$\left(\frac{\partial L(\dot{q})}{\partial \dot{q}} \right)^\top = W\dot{q} + J(q)^\top \lambda = 0 \tag{4.3}$$

$$\left(\frac{\partial L(\dot{q})}{\partial \lambda} \right)^\top = J(q)\,\dot{q} - \dot{r}_{E,d} = 0 \tag{4.4}$$

as a matrix equation

$$\underbrace{\begin{pmatrix} W & J(q)^\top \\ J(q) & O \end{pmatrix}}_{(n+m)\times(n+m)} \begin{pmatrix} \dot{q} \\ \lambda \end{pmatrix} = \begin{pmatrix} 0 \\ \dot{r}_{E,d} \end{pmatrix}$$

shows that a system of $(n+m)$ linear equations must be solved to obtain the vector of joint velocities $\dot{q}$. However, it is possible to reduce the order of the linear system of equations by multiplying the first KKT condition (4.3) with $J(q)\,W^{-1}$, i.e.

$$J(q)\,\underbrace{W^{-1}W}_{E}\,\dot{q} + J(q)\,W^{-1}J(q)^\top \lambda = 0 \tag{4.5}$$

wherein $\mathbf{W}^{-1} = \mathrm{diag}\,(w_1, \dots, w_n)^{-1} = \mathrm{diag}\left(\frac{1}{w_1}, \dots, \frac{1}{w_n}\right)$. Rearranging (4.5) on the previous page and substituting the path-tracking constraint from the second KKT condition (4.4) on the previous page yields

$$\mathbf{J}\,(\mathbf{q})\,\mathbf{W}^{-1}\mathbf{J}\,(\mathbf{q})^{\top}\lambda = -\mathbf{J}\,(\mathbf{q})\,\dot{\mathbf{q}} \quad \rightarrow \quad \mathbf{J}\,(\mathbf{q})\,\mathbf{W}^{-1}\mathbf{J}\,(\mathbf{q})^{\top}\lambda = -\dot{\mathbf{r}}_{\mathrm{E,d}}.$$

One now can derive an expression for the vector of LAGRANGE multipliers

$$\lambda = -\left(\mathbf{J}\,(\mathbf{q})\,\mathbf{W}^{-1}\mathbf{J}\,(\mathbf{q})^{\top}\right)^{-1}\dot{\mathbf{r}}_{\mathrm{E,d}}.$$

In this light, the order of the system of linear equations for $\dot{\mathbf{q}}$ can be reduced to n,

$$\begin{aligned} \dot{\mathbf{q}} &= -\mathbf{W}^{-1}\mathbf{J}\,(\mathbf{q})^{\top}\lambda \\ &= \mathbf{W}^{-1}\mathbf{J}\,(\mathbf{q})^{\top}\underbrace{\left(\mathbf{J}\,(\mathbf{q})\,\mathbf{W}^{-1}\mathbf{J}\,(\mathbf{q})^{\top}\right)^{-1}}_{n \times n}\dot{\mathbf{r}}_{\mathrm{E,d}}. \end{aligned} \tag{4.6}$$

Minimum path-tracking error approach

In this approach from [13] the behavior of the path-tracking error $\mathbf{e}$ is desired to be linear dynamic of single-order and asymptotically stable.

With the definition of the path-tracking error $\mathbf{e}$ and its time derivative $\dot{\mathbf{e}}$

$$\begin{aligned} \mathbf{e} &= \mathbf{r}_{\mathrm{E,d}} - \mathbf{r}_{\mathrm{E}} \\ \dot{\mathbf{e}} &= \dot{\mathbf{r}}_{\mathrm{E,d}} - \dot{\mathbf{r}}_{\mathrm{E}} \\ &= \dot{\mathbf{r}}_{\mathrm{E,d}} - \mathbf{J}\,(\mathbf{q})\,\dot{\mathbf{q}}, \end{aligned}$$

the dynamic behavior of $\mathbf{e}$ is

$$\dot{\mathbf{e}} + \mathbf{K}_{\mathrm{P}}\mathbf{e} = 0$$
$$\left(\dot{\mathbf{r}}_{\mathrm{E,d}} - \mathbf{J}\,(\mathbf{q})\,\dot{\mathbf{q}}\right) + \mathbf{K}_{\mathrm{P}}\mathbf{e} = 0 \tag{4.7}$$

where $\mathbf{K}_{\mathrm{P}} > 0$. Replacing the equality constraint condition in (4.4) on the previous page with (4.7) on the previous page and using $\mathbf{W} = \mathbf{I}$ the KKT conditions are reformulated. With the right MOORE-PENROSE pseudoinverse

$$\mathbf{J}\left(\mathbf{q}\right)^{+} = \mathbf{J}\left(\mathbf{q}\right)^{\top}\left(\mathbf{J}\left(\mathbf{q}\right)\mathbf{J}\left(\mathbf{q}\right)^{\top}\right)^{-1}$$

the vector of joint velocities $\dot{\mathbf{q}}$,

$$\dot{\mathbf{q}} = \mathbf{J}\left(\mathbf{q}\right)^{+}\left(\dot{\mathbf{r}}_{E,d} + \mathbf{K}_P\mathbf{e}\right). \tag{4.8}$$

is obtained.

Remarks on first-order inverse kinematics approaches

The matrix inversions that occurred in course of the preceding derivations are practically computed using numerical methods, especially for larger systems.

The result of the first-order inverse kinematics approaches is the vector of joint velocities $\dot{\mathbf{q}}$. To compute the robot's direct kinematics, this result is integrated by means of numerical methods to obtain the vector of joint positions denoted as $\mathbf{q}$. For longer integration times, inevitably the integration result will drift off the actual solution. In [5] methods for avoiding numerical drift errors are presented. However, in many cases of offline trajectory planning these errors can be neglected due to short simulation times.

Another issue with numerical first-order inverse kinematics approaches occurs when further derivatives of the vector of joint velocities $\dot{\mathbf{q}}$, such as the vector of joint accelerations $\ddot{\mathbf{q}}$, are required for additional calculations such as the computation of the feed-forward torques $\mathbf{Q}\left(\mathbf{q}, \dot{\mathbf{q}}, \ddot{\mathbf{q}}\right)$. Since ideal differential operators,

$$\frac{\mathrm{d}}{\mathrm{d}t} \quad \text{with} \quad \mathcal{L}\left\{\frac{\mathrm{d}}{\mathrm{d}t}\right\} = s,$$

where $\mathcal{L}\left\{\cdot\right\}$ is the LAPLACE transformation, cannot be successfully implemented, a differ-

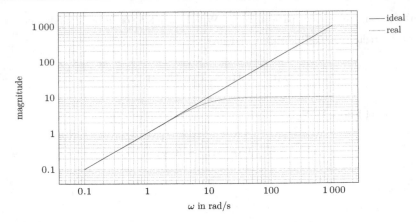

Figure 4.1: Frequency plot of ideal differentiator and differentiation filter with $\omega_0 = 10$

entiation filter represented by its LAPLACE transfer function

$$G(s) = \frac{s}{\frac{s}{\omega_0} + 1}$$

with its cut-off frequency ω_0 is used. The linear state-space model is found to be

$$\dot{x} = -\omega_0 x + u \tag{4.9}$$

$$y = -\omega_0^2 x + \omega_0 u. \tag{4.10}$$

With the output equation for the initial value for the state x, x_0, and the initial value for the input u, u_0, and its time derivative $\dot{u}_0$, i.e.

$$y = \dot{u}_0 = -\omega_0^2 x_0 + \omega_0 u_0$$

follows that

$$x_0 = \frac{\omega_0 u_0 - \dot{u}_0}{\omega_0^2}.$$

Such a replacement element is a high-pass filter with an upper cut-off frequency of ω_0. For signal portions with frequencies above ω_0 the element no longer acts as a differentiator, see the frequency plot in Figure 4.1 on the previous page for a comparison between an ideal and the differentiation filter with $\omega_0 = 10$ rad/s. As a result, such constructs are not favorable for the general case in which fast movements and thus high frequencies may occur, see [13].

4.3.2 Second-order inverse kinematics algorithm

In this section an inverse kinematics approach is described that is based on an asymptotically stable path-tracking error dynamics system of order two.

Using the definition of the path-tracking error $\mathbf{e}$ and its time derivatives $\dot{\mathbf{e}}$ and $\ddot{\mathbf{e}}$, i.e.

$$
\begin{aligned}
\mathbf{e} &= \mathbf{r}_{E,d} - \mathbf{r}_E \\
\dot{\mathbf{e}} &= \dot{\mathbf{r}}_{E,d} - \dot{\mathbf{r}}_E \\
&= \dot{\mathbf{r}}_{E,d} - \mathbf{J}(\mathbf{q})\,\dot{\mathbf{q}} \\
\ddot{\mathbf{e}} &= \ddot{\mathbf{r}}_{E,d} - \ddot{\mathbf{r}}_E \\
&= \ddot{\mathbf{r}}_{E,d} - \dot{\mathbf{J}}(\mathbf{q})\,\dot{\mathbf{q}} - \mathbf{J}(\mathbf{q})\,\ddot{\mathbf{q}},
\end{aligned}
$$

an asymptotically stable behavior of the dynamics of $\mathbf{e}$ is desired to be

$$
\ddot{\mathbf{e}} + \mathbf{K}_D\dot{\mathbf{e}} + \mathbf{K}_P\mathbf{e} = 0
$$
$$
\left(\ddot{\mathbf{r}}_{E,d} - \dot{\mathbf{J}}(\mathbf{q})\,\dot{\mathbf{q}} - \mathbf{J}(\mathbf{q})\,\ddot{\mathbf{q}}\right) + \mathbf{K}_D\dot{\mathbf{e}} + \mathbf{K}_P\mathbf{e} = 0 \tag{4.11}
$$

where $\mathbf{K}_D > 0$ and $\mathbf{K}_P > 0$. Using the second-order error dynamics instead of the single-order form from (4.7) on page 28, similar to (4.8) the dynamic error model (4.11) can be rearranged to obtain the vector of joint accelerations

$$
\ddot{\mathbf{q}} = \mathbf{J}(\mathbf{q})^+ \left(\ddot{\mathbf{r}}_{E,d} - \dot{\mathbf{J}}(\mathbf{q})\,\dot{\mathbf{q}} + \mathbf{K}_D\dot{\mathbf{e}} + \mathbf{K}_P\mathbf{e}\right).
$$

As noted in [12], a drawback of the second-order algorithm is that it is computationally more expensive than first-order algorithms such as those presented in Section 4.3.1 since also the time derivative of the Jacobian, i.e. $\dot{\mathbf{J}}$, is required.

The result obtained from this algorithm is the vector of joint accelerations $\ddot{q}$ that needs to be integrated once to obtain the vector of joint velocities $\dot{q}$ and integrated twice to obtain the vector of joint positions q. Again, for longer integration times, countermeasures against numerical drift errors need to be taken, see [5]. In contrast to velocity-based inverse kinematics approaches, no derivatives with respect to the time t are necessary to be computed for most applications such as the computation of feed-forward actions $Q(q, \dot{q}, \ddot{q})$ of a robot.

4.3.3   Null space projection

While the inverse kinematics algorithms presented in Sections 4.3.1 and 4.3.2 allow to obtain a trajectory for the vector of the manipulator's joint velocities $\dot{q}$ or its vector of joint accelerations $\ddot{q}$, they do not exploit a robot's kinematic redundancy. However, this can be accomplished by augmenting the inverse kinematics solution by an additive term that favors an additional objective, denoted below as the directional vectors $\dot{q}_0$, or $\ddot{q}_0$, respectively.

For first-order inverse kinematics approaches, the solution for the joint velocities $\dot{q}$ can be extended according to

$$\dot{q}^* = \dot{q} + P\dot{q}_0. \tag{4.12}$$

With the choice for the null space projector matrix $P = \left(E - J(q)^+ J(q)\right)$, see [13], the path-tracking constraint (4.2) on page 27 is also satisfied for $\dot{q}^*$,

$$
\begin{aligned}
\dot{r}_{E,d} &= J(q)\dot{q}^* \\
&= J(q)(\dot{q} + P\dot{q}_0) \\
&= J(q)\left(\dot{q} + \left(E - J(q)^+ J(q)\right)\dot{q}_0\right) \\
&= J(q)\dot{q} + \big(\, J(q) - \underbrace{J(q)J(q)^+ J(q)}_{JJ^\top(JJ^\top)^{-1}=E} \,\big)\dot{q}_0 \\
&= J(q)\dot{q}.
\end{aligned}
$$

Similarly, it can be shown for the second-order inverse kinematics approach from Section 4.3.2, that the augmentation of the solution $\ddot{\mathbf{q}}$,

$$\ddot{\mathbf{q}}^* = \ddot{\mathbf{q}} + \mathbf{P}\ddot{\mathbf{q}}_0 \qquad (4.13)$$

wherein again $\mathbf{P} = \left(\mathbf{E} - \mathbf{J}\left(\mathbf{q}\right)^+ \mathbf{J}\left(\mathbf{q}\right)\right)$, meets the path-tracking requirement

$$
\begin{aligned}
\ddot{\mathbf{r}}_{\mathrm{E,d}} &= \dot{\mathbf{J}}\left(\mathbf{q}\right)\dot{\mathbf{q}} + \mathbf{J}\left(\mathbf{q}\right)\ddot{\mathbf{q}}^* \\
&= \dot{\mathbf{J}}\left(\mathbf{q}\right)\dot{\mathbf{q}} + \mathbf{J}\left(\mathbf{q}\right)\left(\ddot{\mathbf{q}} + \mathbf{P}\ddot{\mathbf{q}}_0\right) \\
&= \dot{\mathbf{J}}\left(\mathbf{q}\right)\dot{\mathbf{q}} + \mathbf{J}\left(\mathbf{q}\right)\ddot{\mathbf{q}} + \mathbf{J}\left(\mathbf{q}\right)\mathbf{P}\ddot{\mathbf{q}}_0 \\
&= \dot{\mathbf{J}}\left(\mathbf{q}\right)\dot{\mathbf{q}} + \mathbf{J}\left(\mathbf{q}\right)\ddot{\mathbf{q}} + \mathbf{J}\left(\mathbf{q}\right)\left(\mathbf{E} - \mathbf{J}\left(\mathbf{q}\right)^+ \mathbf{J}\left(\mathbf{q}\right)\right)\ddot{\mathbf{q}}_0 \\
&= \dot{\mathbf{J}}\left(\mathbf{q}\right)\dot{\mathbf{q}} + \mathbf{J}\left(\mathbf{q}\right)\ddot{\mathbf{q}} + \left(\mathbf{J}\left(\mathbf{q}\right) - \underbrace{\mathbf{J}\left(\mathbf{q}\right)\mathbf{J}\left(\mathbf{q}\right)^+ \mathbf{J}\left(\mathbf{q}\right)}_{\mathbf{JJ}^\top(\mathbf{JJ}^\top)^{-1}=\mathbf{E}}\right)\ddot{\mathbf{q}}_0 \\
&= \dot{\mathbf{J}}\left(\mathbf{q}\right)\dot{\mathbf{q}} + \mathbf{J}\left(\mathbf{q}\right)\ddot{\mathbf{q}}.
\end{aligned}
$$

In [13], the author mentions keeping distance from mechanical joint limits, singularities or task space obstacles, or maximizing a manipulability measure as possible objectives for determining the directional vectors $\dot{\mathbf{q}}_0$, or $\ddot{\mathbf{q}}_0$, respectively.

4.3.4 Kinematic manipulability

The following section explains how a robot's kinematic redundancy can be exploited in order to improve a manipulator's property that is called *kinematic manipulability*. Kinematic manipulability describes a robot's ability to change its end-effector velocity depending on its current pose $\mathbf{q}$.

Considering the definition of the ellipsoid of normalized joint velocities

$$\dot{\mathbf{q}}^\top \dot{\mathbf{q}} = 1$$

and substituting the inverse kinematics solution

$$\dot{r}_E = J(q) \dot{q}$$
$$\dot{q} = J(q)^+ \dot{r}_E, \tag{4.14}$$

a task space representation of the ellipsoid can be obtained

$$\dot{r}_E^\top \left(J(q)^+\right)^\top J(q)^+ \dot{r}_E = 1.$$

The kinematic manipulability measure $w_{\text{kin}}(q)$ is defined to be proportional to the volume of the task space velocity ellipsoid,

$$w_{\text{kin}}(q) = \sqrt{\det\left(\left(J(q)^+\right)^\top J(q)^+\right)^{-1}}. \tag{4.15}$$

The joint space gradient of the kinematic manipulability measure, $\frac{\partial w_{\text{kin}}(q)}{\partial q}$, points in the direction of the highest rate of increase of $w_{\text{kin}}(q)$, its length is a measure for the slope of the function in that direction.

The gradient is used to generate a directional vector

$$\dot{q}_0, \ddot{q}_0 = k_w \left(\frac{\partial w_{\text{kin}}(q)}{\partial q}\right)^\top \tag{4.16}$$

that can be utilized to augment the inverse kinematics solutions in order to increase the manipulator's kinematic manipulability while moving along a trajectory, see Section 4.3.3. The scaling factor k_w in (4.16) is used to adjust the vector's length and to convert its physical unit. In this light, one can see that the use of normalized joint velocities $\dot{q}$ can be omitted because scaling only influences the ellipsoid's total size but not its proportions. Thus, appropriate scaling can be accomplished exclusively by means of k_w making any previous normalization obsolete. However, in most literature, such as [5] or [13], the derivation of the kinematic manipulability measure is performed based on normalized joint velocities $\dot{q}$ for reasons of generality. It should be noted that in the above derivations only the end-effector velocity $\dot{r}_E$ but not its orientation are considered for the calculation of the kinematic manipulability as no physical meaning of the determinant expression in (4.15) on the previous page can be established for the simultaneous consideration of both.

Moreover, the kinematic manipulability measure w_{kin} only provides information about a manipulator's general ability to increase its task space velocity $\dot{\mathbf{r}}_E$ in a certain pose $\mathbf{q}$, but does not incorporate any knowledge of a previously specified task space path. Further considerations about this matter can be found in Section 4.3.6.

4.3.5 Dynamic manipulability

The dynamic manipulability of a robot offers insight about its ability to provide end-effector acceleration in task space.

For a manipulator with its equations of motion in matrix form

$$\mathbf{M}\left(\mathbf{q}\right)\ddot{\mathbf{q}} + \mathbf{g}\left(\mathbf{q}, \dot{\mathbf{q}}\right) = \mathbf{Q},$$

the ellipsoid of normalized generalized forces $\mathbf{Q}$ is defined as

$$\mathbf{Q}^{\mathsf{T}}\mathbf{Q} = 1 \qquad (4.17)$$

$$\left(\ddot{\mathbf{q}}^{\mathsf{T}}\mathbf{M}\left(\mathbf{q}\right)^{\mathsf{T}} + \mathbf{g}\left(\mathbf{q}, \dot{\mathbf{q}}\right)^{\mathsf{T}}\right)\left(\mathbf{M}\left(\mathbf{q}\right)\ddot{\mathbf{q}} + \mathbf{g}\left(\mathbf{q}, \dot{\mathbf{q}}\right)\right) = 1$$

$$\ddot{\mathbf{q}}^{\mathsf{T}}\mathbf{M}\left(\mathbf{q}\right)^{\mathsf{T}}\mathbf{M}\left(\mathbf{q}\right)\ddot{\mathbf{q}} + 2\mathbf{g}\left(\mathbf{q}, \dot{\mathbf{q}}\right)^{\mathsf{T}}\mathbf{M}\left(\mathbf{q}\right)\ddot{\mathbf{q}} + \mathbf{g}\left(\mathbf{q}, \dot{\mathbf{q}}\right)^{\mathsf{T}}\mathbf{g}\left(\mathbf{q}, \dot{\mathbf{q}}\right) = 1. \qquad (4.18)$$

In literature, such as [13], a quasi-static approach without consideration of gravitational forces is used, i.e. $\dot{\mathbf{q}} := 0$ and $\mathbf{g}\left(\mathbf{q}, \dot{\mathbf{q}} := 0\right) = 0$. As a result, the ellipsoid equation (4.18) is reduced to a much simpler form

$$\ddot{\mathbf{q}}^{\mathsf{T}}\mathbf{M}\left(\mathbf{q}\right)^{\mathsf{T}}\mathbf{M}\left(\mathbf{q}\right)\ddot{\mathbf{q}} = 1.$$

Using the second-order kinematics relationship with the simplification $\dot{\mathbf{q}} = 0$ applied, i.e.

$$\ddot{\mathbf{r}}_E = \dot{\mathbf{J}}\left(\mathbf{q}\right)\dot{\mathbf{q}} + \mathbf{J}\left(\mathbf{q}\right)\ddot{\mathbf{q}} \qquad (4.19)$$

$$= \mathbf{J}\left(\mathbf{q}\right)\ddot{\mathbf{q}},$$

an inverse kinematics solution can be found by means of the right MOORE-PENROSE pseudoinverse (4.3.1) on page 29, i.e.

$$\ddot{\mathbf{q}} = \mathbf{J}(\mathbf{q})^+ \ddot{\mathbf{r}}_E.$$

Finally, the ellipsoid equation yields

$$\left(\mathbf{J}(\mathbf{q})^+ \ddot{\mathbf{r}}_E\right)^\top \mathbf{M}(\mathbf{q})^\top \mathbf{M}(\mathbf{q})\left(\mathbf{J}(\mathbf{q})^+ \ddot{\mathbf{r}}_E\right) = 1$$
$$\ddot{\mathbf{r}}_E^\top \left(\mathbf{J}(\mathbf{q})^+\right)^\top \mathbf{M}(\mathbf{q})^\top \mathbf{M}(\mathbf{q}) \mathbf{J}(\mathbf{q})^+ \ddot{\mathbf{r}}_E = 1.$$

Without the simplification $\dot{\mathbf{q}} = 0$, the equation for the dynamic manipulability ellipsoid yields

$$\ddot{\mathbf{q}}^\top \mathbf{M}(\mathbf{q})^\top \mathbf{M}(\mathbf{q}) \ddot{\mathbf{q}} = 1 - 2\mathbf{g}(\mathbf{q}, \dot{\mathbf{q}})^\top \mathbf{M}(\mathbf{q}) \ddot{\mathbf{q}} - \mathbf{g}(\mathbf{q}, \dot{\mathbf{q}})^\top \mathbf{g}(\mathbf{q}, \dot{\mathbf{q}}) = A(\mathbf{q}, \dot{\mathbf{q}}, \ddot{\mathbf{q}}).$$

Subsequently, A will be used as a shorthand term for the equation's right hand side. Substituting the inverse form of the kinematics relationship (4.19) on the previous page, i.e.

$$\ddot{\mathbf{q}} = \mathbf{J}(\mathbf{q})^+ \left(\ddot{\mathbf{r}}_E - \dot{\mathbf{J}}(\mathbf{q})\,\dot{\mathbf{q}}\right),$$

the equation for the dynamic manipulability ellipsoid yields

$$\left(\mathbf{J}(\mathbf{q})^+ \left(\ddot{\mathbf{r}}_E - \dot{\mathbf{J}}(\mathbf{q})\,\dot{\mathbf{q}}\right)\right)^\top \mathbf{M}(\mathbf{q})^\top \mathbf{M}(\mathbf{q}) \left(\mathbf{J}(\mathbf{q})^+ \left(\ddot{\mathbf{r}}_E - \dot{\mathbf{J}}(\mathbf{q})\,\dot{\mathbf{q}}\right)\right) = A(\mathbf{q}, \dot{\mathbf{q}}, \ddot{\mathbf{r}}_E).$$

Rearranging this equation yields

$$\ddot{\mathbf{r}}_E^\top (\mathbf{J}^+)^\top \mathbf{M}^\top \mathbf{M} \mathbf{J}^+ \ddot{\mathbf{r}}_E = A - \dot{\mathbf{q}}^\top \dot{\mathbf{J}}^\top (\mathbf{J}^+)^\top \mathbf{M}^\top \mathbf{M} \mathbf{J}^+ \dot{\mathbf{J}} \dot{\mathbf{q}} + 2 \dot{\mathbf{q}}^\top \dot{\mathbf{J}}^\top (\mathbf{J}^+)^\top \mathbf{M}^\top \mathbf{M} \mathbf{J}^+ \ddot{\mathbf{r}}_E$$
$$= B(A, \mathbf{q}, \dot{\mathbf{q}}, \ddot{\mathbf{r}}_E)$$

with the dependency of $\mathbf{q}$ suppressed for the sake of compactness. Substituting A and the second-order inverse kinematics equation (4.19) on page 35, B yields

$$B = 1 - 2\mathbf{g}(\mathbf{q}, \dot{\mathbf{q}})^\top \mathbf{M} \mathbf{J}^+ (\ddot{\mathbf{r}}_E - \dot{\mathbf{J}}\dot{\mathbf{q}}) - \mathbf{g}(\mathbf{q}, \dot{\mathbf{q}})^\top \mathbf{g}(\mathbf{q}, \dot{\mathbf{q}}) +$$
$$- \dot{\mathbf{q}}^\top \dot{\mathbf{J}}^\top (\mathbf{J}^+)^\top \mathbf{M}^\top \mathbf{M} \mathbf{J}^+ \dot{\mathbf{J}}\dot{\mathbf{q}} + 2\dot{\mathbf{q}}^\top \dot{\mathbf{J}}^\top (\mathbf{J}^+)^\top \mathbf{M}^\top \mathbf{M} \mathbf{J}^+ \ddot{\mathbf{r}}_E$$
$$= B(\mathbf{q}, \dot{\mathbf{q}}, \ddot{\mathbf{r}}_E) .$$

Substituting the first-order inverse kinematics equation (4.14) on page 34, reproduced here for quick reference

$$\dot{\mathbf{q}} = \mathbf{J}(\mathbf{q})^+ \dot{\mathbf{r}}_E,$$

yields

$$B = 1 - 2\mathbf{g}(\mathbf{q}, \mathbf{J}^+ \dot{\mathbf{r}}_E)^\top \mathbf{M} \mathbf{J}^+ (\ddot{\mathbf{r}}_E - \dot{\mathbf{J}} \mathbf{J}^+ \dot{\mathbf{r}}_E) - \mathbf{g}(\mathbf{q}, \mathbf{J}^+ \dot{\mathbf{r}}_E)^\top \mathbf{g}(\mathbf{q}, \mathbf{J}^+ \dot{\mathbf{r}}_E) +$$
$$- \dot{\mathbf{r}}_E^\top (\mathbf{J}^+)^\top \dot{\mathbf{J}}^\top (\mathbf{J}^+)^\top \mathbf{M}^\top \mathbf{M} \mathbf{J}^+ \dot{\mathbf{J}} \mathbf{J}^+ \dot{\mathbf{r}}_E + 2\dot{\mathbf{r}}_E^\top (\mathbf{J}^+)^\top \dot{\mathbf{J}}^\top (\mathbf{J}^+)^\top \mathbf{M}^\top \mathbf{M} \mathbf{J}^+ \ddot{\mathbf{r}}_E$$
$$= B(\mathbf{q}, \dot{\mathbf{r}}_E, \ddot{\mathbf{r}}_E)$$

wherein the scalar quantity B is only expressed by means of the robot's current pose $\mathbf{q}$ and the pre-defined path-derivatives $\dot{\mathbf{r}}_E$ and $\ddot{\mathbf{r}}_E$. Alternatively, the second-order inverse kinematics solution (4.3.5) on the previous page can be integrated in order to obtain $\dot{\mathbf{q}}$ which would necessitate an initial velocity value $\dot{\mathbf{q}}(t = 0)$. Consequently, the normalized equation for the dynamic manipulability ellipsoid is found to be

$$\ddot{\mathbf{r}}_E^\top \frac{1}{B(\mathbf{q}, \dot{\mathbf{r}}_E, \ddot{\mathbf{r}}_E)} (\mathbf{J}(\mathbf{q})^+)^\top \mathbf{M}(\mathbf{q})^\top \mathbf{M}(\mathbf{q}) \mathbf{J}(\mathbf{q})^+ \ddot{\mathbf{r}}_E = 1. \tag{4.20}$$

Similarly to the measure for kinematic manipulability, the dynamic manipulability measure $w_{\text{dyn}}(\mathbf{q})$ is defined to be proportional to the volume of the task space acceleration ellipsoid,

$$w_{\text{dyn}}(\mathbf{q}) = \sqrt{\det\left(\frac{1}{B(\mathbf{q}, \dot{\mathbf{r}}_E, \ddot{\mathbf{r}}_E)} (\mathbf{J}(\mathbf{q})^+)^\top \mathbf{M}(\mathbf{q})^\top \mathbf{M}(\mathbf{q}) \mathbf{J}(\mathbf{q})^+ \right)^{-1}}. \tag{4.21}$$

One can see that the initial normalization of the generalized forces $\mathbf{Q}$ in (4.17) on page 35 needs not to be conducted because the second normalization done in (4.20) yields the final ellipsoid equation. In contrast to the derivation of the kinematic manipulability ellipsoid in Section 4.3.4, the second normalization must not be neglected due to the dependence of B on the joint position vector $\mathbf{q}$ of B.

A directional vector $\dot{\mathbf{q}}_0$, or $\ddot{\mathbf{q}}_0$, respectively, that is used for augmenting an inverse kinematics solution from Section 4.3.3 can be obtained from

$$\dot{\mathbf{q}}_0, \ddot{\mathbf{q}}_0 = k_w \left(\frac{\partial w_{\mathrm{dyn}}(\mathbf{q})}{\partial \mathbf{q}} \right)^{\top} \qquad (4.22)$$

wherein the factor k_w is again used adjusting the length and the physical unit. Computing this gradient is recommended to be done by means of numerical methods because of the symbolic complexity of determinant expression from (4.21) on the previous page.

4.3.6   Manipulability along known paths

The manipulability measures w_{kin} and w_{dyn} defined in Sections 4.3.4 and 4.3.5, respectively only provide general information about a robot's ability to change its end-effector velocity $\dot{\mathbf{r}}_{\mathrm{E}}$, or acceleration $\ddot{\mathbf{r}}_{\mathrm{E}}$, respectively. From the sections above follows, that those measures are zero if the end-effector cannot change its velocity or acceleration in any arbitrary task space direction, i.e. the manipulability ellipsoid's volume vanishes which is equivalent to its matrix becoming singular. In this section, a method for obtaining an enhanced manipulability measure that makes use of the knowledge about a predefined path will be presented.

In order to obtain information about the robot's manipulability in the current path direction, one may assume its manipulability ellipsoid placed with its center at the instantaneous point along the path. Following the direction of the current geometric slope of the path, one may find the distance from the center of the ellipsoid to its hull as a manipulability measure.

The example below, which concludes this section will demonstrate how the presented manipulability measure can be explicitly computed.

3D Ellipsoids

In this section, a brief overview of the most important mathematical background of three-dimensional ellipsoids will be presented. Considering a general ellipsoid with its definition

$$_R\mathbf{x}^T\mathbf{V}\,_R\mathbf{x} = 1 \tag{4.23}$$

wherein $_R\mathbf{x}$ denotes the ellipsoid's coordinates in an arbitrary Cartesian system R. Considering the eigenvalue problem of $\mathbf{V} > 0$,

$$\mathbf{V}\mathbf{v}_i = \lambda_i\mathbf{v}_i,$$

wherein λ_i denote the eigenvalues and $\mathbf{v}_i$ are the orthogonal eigenvectors of $\mathbf{V}$. The directions of the principal axes of the ellipsoid are the eigenvectors of $\mathbf{V}$. The lengths a_i of its principal axes are given by the eigenvalues

$$a_i = \sqrt{\frac{1}{\lambda_i}}.$$

In a special three-dimensional case in which the principal axes of the ellipsoid are aligned with the coordinate system in which it is represented, one can obtain the ellipsoid equation (4.23)

$$\frac{\bar{x}^2}{a_1^2} + \frac{\bar{y}^2}{a_2^2} + \frac{\bar{z}^2}{a_3^2} = 1$$

$$\begin{pmatrix} \bar{x} & \bar{y} & \bar{z} \end{pmatrix} \underbrace{\begin{pmatrix} \frac{1}{a_1^2} & 0 & 0 \\ 0 & \frac{1}{a_2^2} & 0 \\ 0 & 0 & \frac{1}{a_3^2} \end{pmatrix}}_{\mathbf{V}} \begin{pmatrix} \bar{x} \\ \bar{y} \\ \bar{z} \end{pmatrix} = 1.$$

Therein the ellipsoid's coordinates are $\bar{x}$, $\bar{y}$ and $\bar{z}$ along its principal axis directions x, y and z. Following [11], the eigenvalues of $\mathbf{V}$ are found to be its diagonal elements, thus the ellipsoid's principal axes lengths are a_1, a_2 and a_3.

2D Example

The following two-dimensional example will help to clarify the above approach of exploiting the knowledge about a pre-defined geometric end-effector path $\mathbf{r}_E$ when enhancing inverse kinematics solutions by increasing the kinematic or dynamic manipulability properties of a manipulator's current pose $\mathbf{q}$. The following considerations can be made for

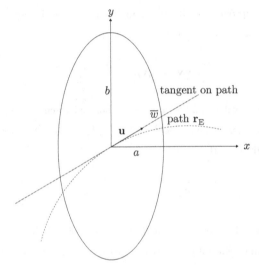

Figure 4.2: Example — manipulability ellipsoid, known path

kinematic and dynamic manipulability ellipsoids equally. Hence starting off of the general two-dimensional ellipsoid from Figure 4.2, i.e. an ellipse, with its definition equation

$$\frac{\overline{x}^2}{a^2} + \frac{\overline{y}^2}{b^2} = 1$$

$$\begin{pmatrix} \overline{x} & \overline{y} \end{pmatrix} \begin{pmatrix} \frac{1}{a^2} & 0 \\ 0 & \frac{1}{b^2} \end{pmatrix} \begin{pmatrix} \overline{x} \\ \overline{y} \end{pmatrix} = 1$$

wherein $\overline{x}$ and $\overline{y}$ are coordinates along the respective coordinate axes x and y and a and b are the lengths of the ellipse's axes.

Figure 4.2 shows the predefined path with the tangential unit vector $\mathbf{u}$ at the current path point r_E, i.e.

$$\mathbf{u} = \begin{pmatrix} u_x \\ u_y \end{pmatrix}, \quad \|\mathbf{u}\| = 1.$$

In general, $\mathbf{u}$ will not be represented in the same coordinate system as the principal axes and requires a coordinate transformation.

Following the direction of $\mathbf{u}$, the distance $\overline{w}$ from the current point along the path to the edge of the ellipse will be used as an augmented manipulabilty measure, subsequently referred to as *directional kinematic manipulability*, or *directional dynamic manipulability*, respectively.

Substituting the ellipse's coordinates $\overline{x}$ and $\overline{y}$ by the elements of a scaled version of the tangential unit vector $\mathbf{u}$

$$\frac{(\overline{w}u_x)^2}{a^2} + \frac{(\overline{w}u_y)^2}{b^2} = 1$$

$$\overline{w}^2 \left(\frac{u_x^2}{a^2} + \frac{u_y^2}{b^2} \right) = 1$$

allows to obtain the *directional manipulability* $\overline{w}$.

$$\overline{w} = \pm \left(\frac{u_x^2}{a^2} + \frac{u_y^2}{b^2} \right)^{-\frac{1}{2}}$$

wherein the negative solution will be neglected since the end-effector moves in positive path direction.

In [13] the author considers $\overline{w}$ as a transformation ratio of the joint space quantities $\mathbf{q}$ or $\mathbf{Q}$ to task space quantities r_E or $\mathbf{F}_E$, respectively.

4.3.7 B-spline-based minimum-time trajectories

A geometric end-effector path r_E can be described as a function of a path parameter variable $s \in [0, 1]$, i.e.

$$r_E = r_E(s).$$

Expressing the path parameter s as a normalized B-spline curve with time t as the curve parameter, i.e.

$$s(t) = \sum_{j=0}^{m-n-2} d_j N_j^n(t), \quad t \in [0, t_E],$$

where m is the number of knots t_k, n is the maximum degree of $s(t)$, t_E is the end time of the trajectory, d_j are the B-spline control points and N_j^n are the B-spline basis functions, yields a reparametrized task space trajectory $r_E(s(t))$.

Now one can find an optimization problem of the form (4.1) on page 24 where the end time of the curve, t_E, and the control points of the curve, d_j, are selected to act as optimization parameters, denoted in vector form as

$$\mathbf{x} = \begin{pmatrix} t_E & d_0 & d_1 & \cdots & d_{m-n-3} & d_{m-n-2} \end{pmatrix}^\top = \begin{pmatrix} t_E & \mathbf{d}_s^\top \end{pmatrix}^\top.$$

If the knots t_k of the B-spline function $s(t)$ are chosen properly, moving the control points d_j allows to locally modify the curve in order to suffice the constraints imposed to the optimization problem.

Additionally, it is possible to assume another B-spline function to be the variation over time of the adjustment factors k_w of the manipulability measure gradients from Sections 4.3.4 and 4.3.5, i.e.

$$\dot{\mathbf{q}}_0, \ddot{\mathbf{q}}_0 = k_{w_{\mathrm{kin}}}(t) \left(\frac{\partial w_{\mathrm{kin}}(\mathbf{q})}{\partial \mathbf{q}} \right)^\top,$$

$$\dot{\mathbf{q}}_0, \ddot{\mathbf{q}}_0 = k_{w_{\mathrm{dyn}}}(t) \left(\frac{\partial w_{\mathrm{dyn}}(\mathbf{q})}{\partial \mathbf{q}} \right)^\top.$$

$k_{w_{\mathrm{kin}}}(t)$ and $k_{w_{\mathrm{dyn}}}(t)$ determine the instantaneous null space contribution to the inverse kinematics solutions from (4.12) on page 32, and (4.13) on page 33, respectively. In an optimization problem of the form (4.1) on page 24 the B-spline control points $d_{\mathrm{kin},i}$ or $d_{\mathrm{dyn},j}$ can be regarded as optimization variables yielding a vector of optimization variables

$$\mathbf{x} = \begin{pmatrix} t_E & \mathbf{d}_s^\top & \mathbf{d}_w^\top \end{pmatrix}^\top \tag{4.24}$$

where $\mathbf{d}_s$ and $\mathbf{d}_w$ are the vectors of control points for the trajectories of the path parameter s and the manipulability measure gradient adjustment factor k_w, respectively. Figure 4.3 shows an overview of this algorithm for the case of a second-order inverse kinematics approach.

This approach enables the inverse kinematics solution to be superimposed by both, kinematic and dynamic manipulability measure gradients simultaneously, i.e.

$$\dot{\mathbf{q}}_0, \ddot{\mathbf{q}}_0 = k_{w_{\mathrm{kin}}}(t) \left(\frac{\partial w_{\mathrm{kin}}(\mathbf{q})}{\partial \mathbf{q}} \right)^{\mathsf{T}} + k_{w_{\mathrm{dyn}}}(t) \left(\frac{\partial w_{\mathrm{dyn}}(\mathbf{q})}{\partial \mathbf{q}} \right)^{\mathsf{T}}$$

yielding a problem of the form (4.1) on page 24 with a vector of optimization variables

$$\mathbf{x} = \left(\begin{array}{cccc} t_{\mathrm{E}} & \mathbf{d}_s^{\mathsf{T}} & \mathbf{d}_{w_{\mathrm{kin}}}^{\mathsf{T}} & \mathbf{d}_{w_{\mathrm{dyn}}}^{\mathsf{T}} \end{array} \right)^{\mathsf{T}}$$

where $\mathbf{d}_{w_{\mathrm{kin}}}$ and $\mathbf{d}_{w_{\mathrm{dyn}}}$ are the control point vectors of the B-spline function trajectories for the adjustment factors. Similar expressions can be found for the manipulability measure for known paths, $\overline{w}$ presented in Section 4.3.6.

Results obtained for the example of a planar three-link manipulator can be found in Section 6.1.

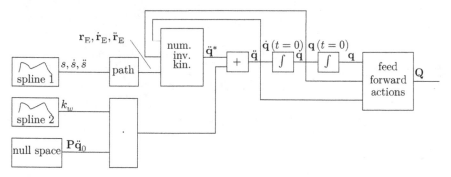

Figure 4.3: B-spline-based minimum-time algorithm — overview

5. Joint Space Decomposition Approach

In this section a separation method for minimum-time trajectory planning for serial redundant robots with one redundant degree of freedom is presented. As seen in Chapter 4, the path in space is parametrized by a scalar parameter s. The approach is based on [9] where the robot joints are divided in two sets, the *non-redundant* of the same dimension as the task space and the remaining *redundant* joint. Time-optimal trajectories for the task space path parameter s and for the *redundant* coordinate q_r are obtained by means of an optimization problem. The *non-redundant* joint positions are found using analytic inverse kinematics for the end-effector position computed by means of the parameterized path and the position of the *redundant* joint.

In the present thesis time-parametrized multi-interval B-spline curves are used expressing the trajectories of the task space path parameter s and the position of the *redundant* degree of freedom q_r. The end time and control points of the curves are used as optimization parameters allowing to incorporate technological and physical constraints of the robot and its environment. Moreover, an approach to reduce the complexity of finding initial values for the optimization process will be presented.

5.1   Method

In Section 4.3 a B-spline curve is used to parametrize the path parameter s of a task space path $r_{E,d}$ and numeric inverse kinematic approaches are used to compute the joint trajectories $q_i(t)$. The present approach adopts the parametrization of the end-effector position path $r_{E,d}$, i.e.

$$r_{E,d} = r_{E,d}(s)$$

with

$$s(t) = \sum_{j=0}^{m_s - n_s - 2} d_{s,j} N_{s,j}^{n_s}(t), \quad t \in [0, t_E],$$

where m_s denotes the number of knots and n_s is the maximum degree of $s(t)$. t_E is the end time of the trajectory, $d_{s,j}$ are the B-spline control points and $N_j^{n_s}$ are the B-spline basis functions.

Similar to [9], the manipulator's joints are separated in one freely selected *redundant* joint q_r and a *non-redundant* part q_{nr} that consists of the rest of the joints, i.e.

$$q_{nr} = S_{nr} q$$

$$q_r = s_r^T q$$

where the selection mask matrix S_{nr} and the selection mask vector s_r perform the separation mentioned above.

The separation process will be further clarified with the following example: If the vector of joint coordinates is $q = \begin{pmatrix} q_1 & q_2 & q_3 \end{pmatrix}^T$ and it is chosen to treat q_3 as the *redundant* degree of freedom, the selection masks S_{nr} and s_r^T are found to be

$$S_{nr} = \text{diag}(1, 1, 0), \quad s_r = \begin{pmatrix} 0 & 0 & 1 \end{pmatrix}^T.$$

For the trajectory of the *redundant* degree of freedom q_r an additional B-spline curve is assumed, i.e.

$$q_r(t) = \sum_{j=0}^{m_r - n_r - 2} d_{r,j} N_{r,j}^{n_r}(t), \quad t \in [0, t_E],$$

where m_r is the number of knots and n_r is the maximum degree of $q_r(t)$. $d_{r,j}$ are the B-spline control points and $N_j^{n_r}$ are the B-spline basis functions. Assuming s and q_r to be known allows to compute one or multiple analytic inverse kinematics solutions for the *non-redundant* degrees of freedom, i.e.

$$\mathbf{q}_{nr} = \mathbf{q}_{nr}\left(\mathbf{r}_{E,d}(s), q_r\right). \tag{5.1}$$

The computation of the joint velocities $\dot{\mathbf{q}}_{nr}$ and the joint acceleration $\ddot{\mathbf{q}}_{nr}$ can be directly performed by deriving (5.1) with respect to time t or by utilizing a velocity-based approach, i.e.

$$\dot{\mathbf{r}}_{E,d} = \mathbf{J}_{nr}\dot{\mathbf{q}}_{nr} \quad \rightarrow \quad \dot{\mathbf{q}}_{nr} = \mathbf{J}_{nr}^{-1}\dot{\mathbf{r}}_{E,d},$$

or an acceleration-based approach, i.e.

$$\ddot{\mathbf{r}}_{E,d} = \dot{\mathbf{J}}_{nr}\dot{\mathbf{q}}_{nr} + \mathbf{J}_{nr}\ddot{\mathbf{q}}_{nr} \quad \rightarrow \quad \ddot{\mathbf{q}}_{nr} = \mathbf{J}_{nr}^{-1}\left(\ddot{\mathbf{r}}_{E,d} - \dot{\mathbf{J}}_{nr}\dot{\mathbf{q}}_{nr}\right),$$

respectively. The the equations above, $\mathbf{J}_{nr}$ denotes the *non-redundant* Jacobian that projects from joint space to task space and $\mathbf{J}_{nr}^+$ is the right MOORE-PENROSE pseudoinverse, see Chapter 4.

This approach necessitates providing an initial trajectory for q_r that guarantees analytic feasibility of the above inverse kinematics problem. In simple cases, such as a path that can be followed by the manipulator with its *redundant* joint locked, one may easily obtain a solution. However, in general, the task of finding a valid initial trajectory for q_r is non-trivial. One possible solution is provided in Section 5.3.

A significant advantage of this separation method is that the resulting trajectories are obtained as polynomial functions that are easy to store and to differentiate whereas the approaches from Sections 4.3.1 and 4.3.2 only yield pointwise trajectories due to the use of numerical inverse kinematics.

5.1.1   Use of B-spline curves

This approach benefits from the adaptability of B-spline curves to the requirements of
the trajectory. Due to the B-spline property of local approximation, see Section 2.1.1,
this class of functions is especially suited for optimization tasks, because changes of one
control point in the vector of optimization parameters $\mathbf{x}$ only produces a local change
of the resulting trajectory. Additionally, by adjusting the position and the density of
the knots where rapid changes are expected, the function's versatility can be improved.
Figure 5.1 depicts an overview of the B-spline-based separation algorithm.

It is also possible to specify certain initial or final conditions for s and q_r by arranging
the first or last control points accordingly.

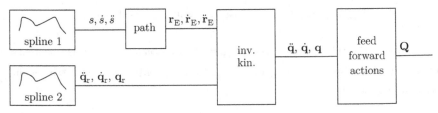

Figure 5.1: B-spline-based separation algorithm — overview

5.2   Optimization problem

The separation method presented in this section is applied by means of an optimization
problem similar to the one defined in Section 4.1. The goal is to minimize the trajectory
end-time t_E, i.e.

$$\mathbf{x}^* = \arg\min_{\mathbf{x}} f(\mathbf{x}) = \arg\min_{\mathbf{x}} t_E(\mathbf{x}),$$
$$\text{s.t.} \quad \mathbf{c}_{eq} = \mathbf{0}$$
$$\mathbf{c}_{ineq} \le \mathbf{0}$$

The path-following equality constraint for the task space path

$$\mathbf{r}_{\mathrm{E,d}}\left(s\right), \quad s = s\left(t\right) \in [0,1], \quad t \in [0,t_{\mathrm{E}}], \quad s\left(t = t_{\mathrm{E}}\right) = 1,$$

is incorporated by the analytical inverse kinematics approach described in the previous section.

Examples for inequality constraints are constrained joint velocities

$$\mathbf{c}_{\mathrm{ineq},\dot{\mathbf{q}}}\left(\mathbf{x}\right) = \left(\begin{array}{c} \dot{\mathbf{q}} - \dot{\mathbf{q}}_{\mathrm{max}} \\ \dot{\mathbf{q}}_{\mathrm{min}} - \dot{\mathbf{q}} \end{array} \right) \leq \mathbf{0},$$

or joint accelerations

$$\mathbf{c}_{\mathrm{ineq},\ddot{\mathbf{q}}}\left(\mathbf{x}\right) = \left(\begin{array}{c} \ddot{\mathbf{q}} - \ddot{\mathbf{q}}_{\mathrm{max}} \\ \ddot{\mathbf{q}}_{\mathrm{min}} - \ddot{\mathbf{q}} \end{array} \right) \leq \mathbf{0},$$

or constrained input actions

$$\mathbf{c}_{\mathrm{ineq},\mathbf{Q}}\left(\mathbf{x}\right) = \left(\begin{array}{c} \mathbf{Q} - \mathbf{Q}_{\mathrm{max}} \\ \mathbf{Q}_{\mathrm{min}} - \mathbf{Q} \end{array} \right) \leq \mathbf{0},$$

where the argument $\mathbf{x}$ again was omitted in the subvectors. $\mathbf{q}$ denotes the vector of joint positions, $\dot{\mathbf{q}}$ are the joint velocities, $\ddot{\mathbf{q}}$ are the joint accelerations. The input actions $\mathbf{Q}$ are computed by means of inverse dynamics $\mathbf{Q} = \mathbf{Q}\left(s, \dot{s}, \ddot{s}, q_{\mathrm{r}}, \dot{q}_{\mathrm{r}}, \ddot{q}_{\mathrm{r}}\right)$ requiring the equations of motions of the considered robot system.

In contrast to Section 4.1, here the vector of optimization variables $\mathbf{x}$ is

$$\mathbf{x} = \left(\begin{array}{ccc} t_{\mathrm{E}} & \mathbf{d}_{s}^{\top} & \mathbf{d}_{\mathrm{r}}^{\top} \end{array} \right)^{\top}$$

where $\mathbf{d}_{s}$ and $\mathbf{d}_{\mathrm{r}}$ are the vectors of control points of the B-spline curves for the path parameter s and the redundant joint position q_{r}, respectively.

This optimization problem can be solved by means of the *Active Set* method mentioned in Section 4.2.

5.3 Initial trajectories

Regarding the task of finding an initial trajectory that provides geometric feasibility to the inverse kinematics problem in the previous section, this segment features one possible solution method.

As a first step, the trajectory with end time $t_{E,0}$ for the path parameter s is determined. Since $s \in [0,1]$ and $n_s - 1$ derivatives of s with respect to time t are zero at the beginning and the end of the trajectory, the n_s initial control points are set to 0 and the final n_s control points are set to 1 while the remaining control points in the middle are equally spaced between 0 and 1, i.e.

$$\left(\underbrace{0 \; \cdots \; 0}_{n_s} \; \mathbf{d}_{s,0}^{\top} \; \underbrace{1 \; \cdots \; 1}_{n_s} \right)^{\top}. \tag{5.2}$$

Next, an inverse kinematics approach from section 4.3.1 can be used to compute a solution for the joint velocities $\dot{q}_i(t)$ of the end-effector trajectory generated using (5.2). Integration of the velocity of the *redundant* joint, $\dot{q}_r$, yields the joint position $q_r(t)$.

As shown in Section 2.1.3, this trajectory can then be approximated with a B-spline of maximum degree n_r with m_r knots, resulting in $m_r - n_r - 1$ control points $\mathbf{d}_{r,0}$, see [7].

The parameterizations of s and q_r can then be used as initial values

$$\left(t_{E,0} \; \mathbf{d}_{s,0}^{\top} \; \mathbf{d}_{r,0}^{\top} \right)^{\top}$$

for the optimization problem from Section 5.2.

5.4 Remaining challenges

As the previous sections show, this approach depends on a number of parameters such as the selection of the separated, *redundant* joint, the number and interval sizes of knots, the number of control points, the polynomial degree of the B-spline curves for s and q_r, the initial pose of the robot and the initial values for the optimization process as well as the choice of the inverse kinematics solution. Thus, only a local minimum of the end-time can be found using a single set of parameters. In a multi-level optimization problem that also varies above parameters, improved solutions may be achieved. As the present formulation not necessarily yields a convex optimization problem, the result is not a global minimum in general.

While the polynomial degree is practically chosen to meet the continuity requirements for the trajectory, it must be at least continuously differentiable twice for the computation of the Hessian matrix in course of the applied SQP-like optimization method. If it is also chosen to incorporate inequality constraints of the continous input actions, the second order derivatives $\ddot{q}$ need to be computed by means of inverse dynamics. The number of initial parameters for the optimization process can be significantly reduced by means of the approach presented in Section 5.3, the rest is determined empirically.

For covering the task of finding minimum-time trajectories for serial robots that feature more than one redundant degree of freedom, separate B-spline trajectories can be assumed for each redundant degree of freedom. This adds additional complexity to the task of determining geometrically feasible trajectories as initial values for the optimization process.

Furthermore, the separation approach relies on analytic inverse kinematics solutions that are mostly derived from the manipulator's geometric properties. Inverse kinematics is the injective transformation from task space to joint space, i.e. in general there can be multiple solutions for joint configurations that yield the same end-effector position. For general end-effector paths, it may be required to use different inverse kinematics solutions for different parts of the path to provide feasibility with respect to the physical limits of the manipulator. Hence, methods that allow to switch between distinct solutions during the path are subject to further investigation.

6. Examples

6.1 Planar robot

In this section, the separation method presented in Chapter 5 will be used to generate minimum-time trajectories for a selective compliance articulated robot arm (SCARA) with three degrees of freedom. Examples for applications of robots of this type are assembly tasks or pick-and-place operations in industrial environments. Introducing minimum-time trajectories to these tasks, reduced operation cycle times and thus increased cost-effectiveness can be achieved.

In this example, only the end-effector position, but not its orientation is considered for path-planning. Hence the vector of end-effector coordinates yields

$$_I\mathbf{z}_E = {_I\mathbf{r}_E} = \begin{pmatrix} x \\ y \end{pmatrix}_I.$$

For principal considerations, it is sufficient to assume that the links of the manipulator lie within a plane, see Figure 6.1 on the next page. First, the forward kinematics, i.e. the transformation from joint space to task space, is obtained from the manipulator's geometric properties. Then, three solution paths to inverse kinematics, i.e. the transformation from task space to joint space, are derived by means of basic trigonometry. As described in Chapter 5, in each case one of the robot's three joints is treated as redundant and thus is assumed to be known for the analytic computation of the inverse kinematics solution,

53

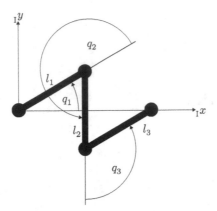

Figure 6.1: Planar robot

hence the need for three approaches to inverse kinematics. Furthermore, a dynamic model of the manipulator in the form of the equations of motion is derived using the *Projection Equation*, a synthetic method presented in [2]. Finally, optimization results for straight line paths are presented and compared against results obtained using methods derived in Section 4.3 as well as results given in [9].

6.1.1   Forward kinematics

End-effector position

The task space coordinate representation of the end-effector position can be easily derived from Figure 6.1,

$$
{}_I\mathbf{r}_E = \begin{pmatrix} x_E \\ y_E \end{pmatrix}_I = \begin{pmatrix} l_1 \cos(q_1) + l_2 \cos(q_1 + q_2) + l_3 \cos(q_1 + q_2 + q_3) \\ l_1 \sin(q_1) + l_2 \sin(q_1 + q_2) + l_3 \sin(q_1 + q_2 + q_3) \end{pmatrix}_I . \tag{6.1}
$$

End-effector velocity, Jacobian

The task space velocity $\dot{\mathbf{r}}_E$ is obtained by differentiating (6.1) with respect to time t after replacing the joint space coordinates $q_i, i \in \{1, 2, 3\}$ with the respective time-dependent functions $q_i(t)$,

$$_I\dot{\mathbf{r}}_E = _I\begin{pmatrix} -l_1 \sin(q_1)\dot{q}_1 - l_2 \sin(q_1+q_2)(\dot{q}_1+\dot{q}_2) - l_3 \sin(q_1+q_2+q_3)(\dot{q}_1+\dot{q}_2+\dot{q}_3) \\ l_1 \cos(q_1)\dot{q}_1 + l_2 \cos(q_1+q_2)(\dot{q}_1+\dot{q}_2) + l_3 \cos(q_1+q_2+q_3)(\dot{q}_1+\dot{q}_2+\dot{q}_3) \end{pmatrix}. \tag{6.2}$$

Rewriting (6.2) as a matrix equation,

$$_I\dot{\mathbf{r}}_E = \begin{pmatrix} -l_1 \sin(q_1) - l_2 \sin(q_1+q_2) - l_3 \sin(q_1+q_2+q_3) & l_1 \cos(q_1) + l_2 \cos(q_1+q_2) + l_3 \cos(q_1+q_2+q_3) \\ -l_2 \sin(q_1+q_2) - l_3 \sin(q_1+q_2+q_3) & l_2 \cos(q_1+q_2) + l_3 \cos(q_1+q_2+q_3) \\ -l_3 \sin(q_1+q_2+q_3) & l_3 \cos(q_1+q_2+q_3) \end{pmatrix}^\top \begin{pmatrix} \dot{q}_1 \\ \dot{q}_2 \\ \dot{q}_3 \end{pmatrix}$$

$$= \mathbf{J}(\mathbf{q})\,\dot{\mathbf{q}},$$

one can obtain the Jacobian matrix $\mathbf{J}(\mathbf{q})$ of the task space velocities $_I\dot{\mathbf{r}}_E$ with respect to the joint space velocities $\dot{\mathbf{q}}$. However, this matrix representation depends on the selection of the coordinate frame in which the task space velocities are represented, see [13]. As the Jacobian is always derived from inertial velocities, any indication of a frame of reference will be omitted. More general forms of the Jacobian, the geometric and analytic Jacobian, additionally incorporate the end-effector orientation change rate, see [5] or [12]. In a simple case where only the end-effector position is considered those forms are equivalent.

End-effector acceleration

The task space acceleration $\ddot{\mathbf{r}}_E$ can be obtained by differentiating (6.2) with respect to time t after replacing the coordinates q_i and $\dot{q}_i$ with the time-dependent functions $q_i(t)$ and $\dot{q}_i(t)$, respectively,

$$_I\ddot{\mathbf{r}}_E = \frac{\mathrm{d}}{\mathrm{d}t}{}_I\dot{\mathbf{r}}_E$$

$$= \dot{\mathbf{J}}(\mathbf{q})\,\dot{\mathbf{q}} + \mathbf{J}(\mathbf{q})\,\ddot{\mathbf{q}}.$$

Explicit results are omitted for the sake of brevity as they will not be needed for the subsequent derivations.

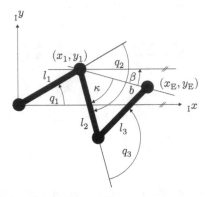

Figure 6.2: Planar robot — Inverse kinematics $q_r = q_1$

6.1.2 Inverse kinematics $q_r = q_1$

For finding this inverse kinematics solution, see Figure 6.2, the joint angle q_1 is assumed to be known in terms of the separation method presented in this thesis. Hence, the position of the end point of link 1 is known,

$$_I\mathbf{r}_1 = \begin{pmatrix} x_1 \\ y_1 \end{pmatrix}_I = \begin{pmatrix} l_1 \cos(q_1) \\ l_1 \sin(q_1) \end{pmatrix}_I$$

and the problem can be reduced to the inverse kinematics of a two-link planar manipulator. Considering the range from the origin to the end-effector position

$$l = \sqrt{x_E^2 + y_E^2},$$

and the distance from the end point of link 1 to the end-effector position

$$b = \sqrt{(x_E - x_1)^2 + (y_E - y_1)^2}$$

and the angle β,

$$\beta = \arctan(y_E - y_1, x_E - x_1),$$

the auxiliary angle κ can be determined,

$$\kappa = \pm \arccos \left(\frac{b^2 + l_2^2 - l_3^2}{2\,b\,l_2} \right).$$

Now it is possible to find an expression for the joint angle q_2,

$$q_2 = \kappa - \beta - q_1.$$

From the second component of (6.1) on page 54, reproduced here for quick reference,

$$y_{\mathrm{E}} = l_1 \sin(q_1) + l_2 \sin(q_1 + q_2) + l_3 \sin(q_1 + q_2 + q_3),$$

follows that

$$q_3 = \arcsin \left(\frac{y_{\mathrm{E}} - l_1 \sin(q_1) - l_2 \sin(q_1 + q_2)}{l_3} \right) - q_1 - q_2.$$

6.1.3 Inverse kinematics $q_{\mathrm{r}} = q_2$

Following the separation method presented in this thesis, the joint angle q_2 is treated as redundant and is assumed to be known for the following calculations. Consequently, the connected links 1 and 2 can be equivalently represented by one single link, subsequently referred to as *virtual link*, of length a and internal angle β,

$$\begin{aligned} a &= \sqrt{(l_1 + l_2 \cos(q_2))^2 + (l_2 \sin(q_2))^2} \\ \beta &= \arctan(l_2 \sin(q_2), l_1 + l_2 \cos(q_2)), \end{aligned}$$

see also Figure 6.3 on the next page.

Figure 6.4 on page 59 shows a general configuration of the manipulator with the resulting *virtual link* which allows to derive a set of inverse kinematics equations for the joint angles q_1 and q_3. From the transformation of the cartesian representation of the end-effector

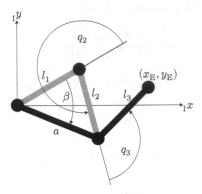

Figure 6.3: Planar robot — Inverse kinematics $q_r = q_2$, virtual link

coordinates to polar coordinates range l and bearing α,

$$l = \sqrt{x_E^2 + y_E^2}$$
$$\alpha = \arctan(y_E, x_E),$$

an expression for the auxiliary angle κ can be found,

$$\kappa = \pm \arccos \left(\frac{l^2 + a^2 - l_3^2}{2\,l\,a} \right),$$

considering that the cosine is an even function. From a comparison of Figure 6.3 and Figure 6.4 on the next page, the joint angle q_1 can be determined,

$$q_1 = \alpha + \kappa - \beta.$$

In Figure 6.3 it can be seen that the remaining joint angle q_3 is

$$q_3 = \arctan\left(y_E - l_1 \sin(q_1) - l_2 \sin(q_1 + q_2),\, x_E - l_1 \cos(q_1) - l_2 \cos(q_1 + q_2)\right) - q_1 - q_2.$$

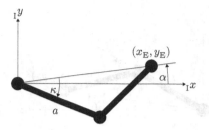

Figure 6.4: Planar robot — Inverse kinematics $q_r = q_2$

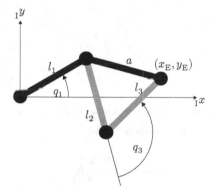

Figure 6.5: Planar robot — Inverse kinematics $q_r = q_3$, virtual link

6.1.4 Inverse kinematics $q_r = q_3$

Due to the assumption that q_3 is known, the adjoined links 2 and 3 can be reduced to one single, *virtual link* of length a,

$$a = \sqrt{(l_2 + l_3 \cos(q_3))^2 + (l_3 \sin(q_3))^2},\tag{6.3}$$

as depicted in Figure 6.5. After transforming the cartesian representation of the end-effector position to the polar form, range l and bearing α,

$$l = \sqrt{x_E^2 + y_E^2}\tag{6.4}$$

$$\alpha = \arctan(y_E, x_E),\tag{6.5}$$

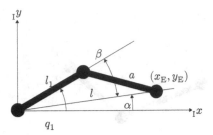

Figure 6.6: Planar robot — Inverse kinematics $q_r = q_3$ (1)

and applying the law of cosines to the triangle with edges a, l and l_1 from Figure 6.6, an expression for the auxiliary angle β can be found,

$$\beta = \pm \arccos \left(\frac{l^2 + l_1^2 - a^2}{2\,l\,l_1} \right),$$

considering that the cosine is an even function. Now the joint angle q_1 can be computed,

$$q_1 = \alpha - \beta.$$

With the auxiliary angles γ and κ from Figure 6.7 on the next page,

$$\begin{aligned}
\gamma &= \arctan\left(l_3 \sin(q_3),\, l_2 + l_3 \cos(q_3) \right) \\
\kappa &= \arctan\left(y_E - l_1 \sin(q_1),\, x_E - l_1 \cos(q_1) \right),
\end{aligned}$$

one can find an expression for the remaining joint angle q_2,

$$q_2 = \kappa - q_1 - \gamma.$$

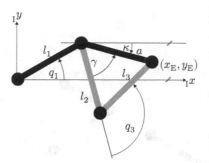

Figure 6.7: Planar robot — Inverse kinematics $q_r = q_3$ (2)

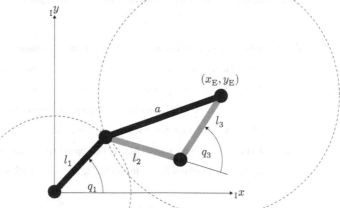

Figure 6.8: Planar robot — Singularity avoidance, feasible configuration

6.1.5 Avoidance of singularities due to choice of inverse kinematics approach

In each of the inverse kinematics approaches presented in Sections 6.1.2 to 6.1.4, one joint angle is treated as redundant, i.e. assumed to be known, as it is determined separately in terms of the presented separation method.

The following example will show that the redundant joint angle q_r may not assume arbitrary values for a given end-effector position but that an admissible interval can be determined under certain conditions. In Figure 6.8 the third joint is treated as redundant, i.e. $q_r = q_3$. As described in Section 6.1.4, links 2 and 3 can be reduced to a single, *virtual link* with length a. From Figure 6.9 on the next page one can see that it is geomet-

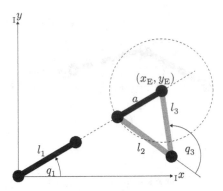

Figure 6.9: Planar robot — Singularity avoidance, infeasible configuration

rically not feasible to find an inverse kinematics solution for the suggested choice of q_3, since no connection point for the first link and the virtual link can be found. Obviously, the feasibility of the inverse kinematics problem is bounded in the pose where the *virtual link* and the remaining link 1 assume a singular configuration. From the configuration depicted in Figure 6.8 on the previous page it can be seen that such a singularity will occur if the length of the *virtual link*, $a = a\,(q_3)$, is decreased to the difference between the distance of the end-effector position from the origin, l, and the length of link 1, l_1, i.e.

$$a_{\mathrm{sing}} = l - l_1.$$

From rearranging (6.3) on page 59,

$$q_{3,\mathrm{sing}} = \pm \arccos\left(\frac{a_{\mathrm{sing}}^2 - l_2^2 - l_3^2}{2\,l_2\,l_3}\right),$$

bounds for q_3 can be obtained at which a singularity of the *virtual link* and link 1 will occur.

According to the procedure for obtaining bounds for q_3 above, bounds for q_2 can be found for the inverse kinematics case $q_{\mathrm{r}} = q_2$ as described in Section 6.1.3. Geometric considerations of Figure 6.3 on page 58 and Figure 6.4 on page 59 yield the length of the *virtual link* consisting of links 1 and 2 at a singular configuration of the *virtual link* with link 3,

$$a_{\text{sing}} = l - l_3.$$

It can be seen that the bounds for q_2 are

$$q_{2,\text{sing}} = \pm \arccos \left(\frac{a_{\text{sing}}^2 - l_1^2 - l_2^2}{2 \, l_1 \, l_2} \right),$$

Similarly, bounds for q_1 can be computed in the inverse kinematics approach from Section 6.1.2 where the joint angle q_1 is being treated as redundant, i.e. $q_r = q_1$. From Figure 6.2 on page 56 it can be seen, that the manipulator links 2 and 3 will assume a singular configuration when q_1 reaches its boundary values

$$q_{1,\text{sing}} = \alpha \pm \arccos \left(\frac{(l_2 + l_3)^2 - l_1^2 - l^2}{2 \, l_1 \, l} \right),$$

where $\alpha = \arctan (y_{\text{E}}, x_{\text{E}})$.

6.1.6 Kinematic chain

Before the dynamic system is derived in Section 6.1.7 using the subsystem formulation of the *Projection equation* that is presented in Chapter 3, the kinematic relationship between the subsystems needs to be established.

Rotation matrices

First, the rotational coordinate transformations between the subsystem coordinate frames depicted in Figure 6.10 on the next page are derived.

$$\mathbf{A}_{12} = \begin{pmatrix} \cos(q_2) & -\sin(q_2) & 0 \\ \sin(q_2) & \cos(q_2) & 0 \\ 0 & 0 & 1 \end{pmatrix},$$

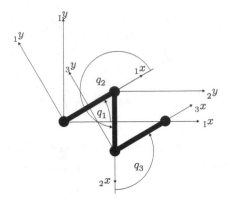

Figure 6.10: Planar robot — Coordinate frames

$$\mathbf{A}_{23} = \begin{pmatrix} \cos(q_3) & -\sin(q_3) & 0 \\ \sin(q_3) & \cos(q_3) & 0 \\ 0 & 0 & 1 \end{pmatrix}.$$

Kinematic chain

Following (3.3) on page 18, the kinematic relationship between the subsystems can be expressed as:

$$\begin{aligned}
\dot{\mathbf{y}}_1 &= \mathbf{f}_1 \dot{s}_1 \\
&= \begin{pmatrix} 0 & 0 & 0 & 0 & 0 & 0 & 1 \end{pmatrix}^\top \dot{q}_1 \\
&= \begin{pmatrix} 0 & 0 & 0 & 0 & 0 & 0 & \dot{q}_1 \end{pmatrix}^\top,
\end{aligned}$$

$$\begin{aligned}
\dot{\mathbf{y}}_2 &= \mathbf{T}_{2,1} \dot{\mathbf{y}}_1 + \mathbf{f}_2 \dot{s}_2 \\
&= \mathbf{T}_{2,1} \dot{\mathbf{y}}_1 + \begin{pmatrix} 0 & 0 & 0 & 0 & 0 & 0 & 1 \end{pmatrix}^\top \dot{q}_2 \\
&= \mathbf{T}_{2,1} \dot{\mathbf{y}}_1 + \begin{pmatrix} 0 & 0 & 0 & 0 & 0 & 0 & \dot{q}_2 \end{pmatrix}^\top,
\end{aligned}$$

where only the essential column $\mathbf{f}_1$ of the functional Matrix $\mathbf{F}_1$ is used and a reduced form of the transformation matrix is

$$
\mathbf{T}_{2,1} = \begin{pmatrix} \mathbf{A}_{21} & \mathbf{A}_{211}\tilde{\mathbf{r}}_{12}^{\mathsf{T}} & \mathbf{A}_{211}\tilde{\mathbf{r}}_{12}^{\mathsf{T}}\mathbf{e}_3 \\ \mathbf{O} & \mathbf{A}_{21} & \mathbf{A}_{21}\mathbf{e}_3 \\ 0\ 0\ 0 & 0\ 0\ 0 & 0 \end{pmatrix}
$$

with ${}_1\mathbf{r}_{12} = \begin{pmatrix} l_1 & 0 & 0 \end{pmatrix}^{\mathsf{T}}$. The auxiliary velocities of the third subsystems are found to be

$$
\begin{aligned}
\dot{\mathbf{y}}_3 &= \mathbf{T}_{3,2}\dot{\mathbf{y}}_2 + \mathbf{f}_3\dot{s}_3 \\
&= \mathbf{T}_{3,2}\dot{\mathbf{y}}_2 + \begin{pmatrix} 0 & 0 & 0 & 0 & 0 & 0 & 1 \end{pmatrix}^{\mathsf{T}}\dot{q}_3 \\
&= \mathbf{T}_{3,2}\dot{\mathbf{y}}_2 + \begin{pmatrix} 0 & 0 & 0 & 0 & 0 & 0 & \dot{q}_3 \end{pmatrix}^{\mathsf{T}}
\end{aligned}
$$

with a reduced form of the transformation matrix

$$
\mathbf{T}_{3,2} = \begin{pmatrix} \mathbf{A}_{32} & \mathbf{A}_{322}\tilde{\mathbf{r}}_{23}^{\mathsf{T}} & \mathbf{A}_{322}\tilde{\mathbf{r}}_{23}^{\mathsf{T}}\mathbf{e}_3 \\ \mathbf{O} & \mathbf{A}_{32} & \mathbf{A}_{32}\mathbf{e}_3 \\ 0\ 0\ 0 & 0\ 0\ 0 & 0 \end{pmatrix}
$$

and ${}_2\mathbf{r}_{23} = \begin{pmatrix} l_2 & 0 & 0 \end{pmatrix}^{\mathsf{T}}$. In the light of (3.2) on page 17 and the derivations above, the synthesis of the equations of motion can be completed using the full functional matrices

$$
\mathbf{F}_1 = \left(\frac{\partial \dot{\mathbf{y}}_1}{\partial \dot{s}}\right)^{\mathsf{T}} = \begin{pmatrix} 0 & 0 & 0 & 0 & 0 & 0 & 1 \\ 0 & 0 & 0 & 0 & 0 & 0 & 0 \\ 0 & 0 & 0 & 0 & 0 & 0 & 0 \end{pmatrix}^{\mathsf{T}}
$$

$$
\mathbf{F}_2 = \left(\frac{\partial \dot{\mathbf{y}}_2}{\partial \dot{s}}\right)^{\mathsf{T}} = \begin{pmatrix} l_1\sin(q_2) & l_1\cos(q_2) & 0 & 0 & 0 & 1 & 0 \\ 0 & 0 & 0 & 0 & 0 & 0 & 1 \\ 0 & 0 & 0 & 0 & 0 & 0 & 0 \end{pmatrix}^{\mathsf{T}}
$$

$$\mathbf{F}_3 = \left(\frac{\partial \dot{\mathbf{y}}_3}{\partial \dot{\mathbf{s}}}\right)^{\mathsf{T}}$$

$$= \begin{pmatrix} l_1 \sin\left(q_2 + q_3\right) + l_2 \sin\left(q_3\right) & l_1 \cos\left(q_2 + q_3\right) + l_2 \cos\left(q_3\right) & 0 & 0 & 0 & 1 & 0 \\ l_2 \sin\left(q_3\right) & l_2 \cos\left(q_3\right) & 0 & 0 & 0 & 1 & 0 \\ 0 & 0 & 0 & 0 & 0 & 0 & 1 \end{pmatrix}^{\mathsf{T}}.$$

6.1.7 Dynamic model

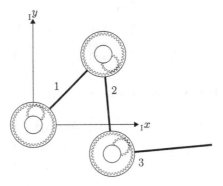

Figure 6.11: Planar robot consisting of three subsystems

Analysis of the manipulator's characteristics shows that it consists of a series of three subsystems with identical structure, i.e. a link that is connected to a motor via gears, see Figure 3.1 on page 18. It is practical to choose a modular approach by creating a dynamic model of the subsystem first and then synthesizing the entire model. In Chapter 3, the *Projection Equation* is derived in a formulation suitable for the present subsystem class.

Synthesis of subsystems

As the kinematics have been covered in the previous section and the subsystem matrices for the dynamic system can be obtained from Chapter 3, only the respective predecessor velocities and the relevant model parameters need to be substituted,

$$\omega_{0,1} = \begin{pmatrix} 0 \\ 0 \\ \dot{q}_1 \end{pmatrix}, \quad \omega_{\text{arm}} = \dot{q}_1,$$

$$\begin{aligned}
m_{\text{arm}} &= m_{\text{arm1}} \\
C_{\text{c,arm}} &= C_{\text{c,arm1}} \\
s_{\text{arm}} &= s_{\text{arm1}} \\
m_{\text{mot}} &= m_{\text{mot1}} \\
C_{\text{c,mot}} &= C_{\text{c,mot1}} \\
i_{\text{G}} &= i_{\text{G},1} \\
M_{\text{mot}} &= M_{\text{mot},1} \\
d &= d_1
\end{aligned}$$

$$\omega_{0,2} = \begin{pmatrix} 0 \\ 0 \\ \dot{q}_1 + \dot{q}_2 \end{pmatrix}, \quad \omega_{\text{arm}} = \dot{q}_2,$$

$$\begin{aligned}
m_{\text{arm}} &= m_{\text{arm2}} \\
C_{\text{c,arm}} &= C_{\text{c,arm2}} \\
s_{\text{arm}} &= s_{\text{arm2}} \\
m_{\text{mot}} &= m_{\text{mot2}} \\
C_{\text{c,mot}} &= C_{\text{c,mot2}} \\
i_{\text{G}} &= i_{\text{G},2} \\
M_{\text{mot}} &= M_{\text{mot},2} \\
d &= d_2
\end{aligned}$$

$$\omega_{0,3} = \begin{pmatrix} 0 \\ 0 \\ \dot{q}_1 + \dot{q}_2 + \dot{q}_3 \end{pmatrix}, \quad \omega_{\text{arm}} = \dot{q}_3,$$

$$\begin{aligned}
m_{\text{arm}} &= m_{\text{arm3}} \\
C_{\text{c,arm}} &= C_{\text{c,arm3}} \\
s_{\text{arm}} &= s_{\text{arm3}} \\
m_{\text{mot}} &= m_{\text{mot3}} \\
C_{\text{c,mot}} &= C_{\text{c,mot3}} \\
i_{\text{G}} &= i_{\text{G},3} \\
M_{\text{mot}} &= M_{\text{mot},3} \\
d &= d_3.
\end{aligned}$$

6.1.8   Optimization Results

Task definition

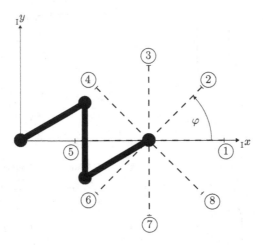

Figure 6.12: Planar robot — Paths for tasks

The task for this example is to find minimum-time trajectories for the manipulator from Figure 6.1 on page 54 moving along the straight line end-effector paths from the initial pose

$$q_1\,(t=0) = \frac{1}{6}\pi, \quad q_2\,(t=0) = -\frac{2}{3}\pi, \quad q_3\,(t=0) = \frac{2}{3}\pi$$

drawn in Figure 6.12 to the end points 1 to 8. The physical parameters that were used to derive the kinematic and the dynamic model of the robot in Sections 6.1.1 to 6.1.7 are

$$m_{\mathrm{arm1}} = m_{\mathrm{arm2}} = m_{\mathrm{arm3}} = m_{\mathrm{arm}} = 10 \text{ kg}$$

$$l_1 = l_2 = l_3 = l = 1 \text{ m}$$

$$C_{\mathrm{c,arm1}} = C_{\mathrm{c,arm2}} = C_{\mathrm{c,arm3}} = m_{\mathrm{arm}}\frac{l^2}{12} = 0.8333 \text{ kgm}^2$$

$$s_{\mathrm{arm1}} = s_{\mathrm{arm2}} = s_{\mathrm{arm3}} = s_{\mathrm{arm}} = \frac{l}{2} = 0.5 \text{ m}$$

$$m_{\text{mot1}} = m_{\text{mot2}} = m_{\text{mot3}} = m_{\text{mot}} = 0$$

$$i_{G,1} = i_{G,2} = i_{G,3} = i_G = 1$$

$$C_{c,\text{mot1}} = C_{c,\text{mot2}} = C_{c,\text{mot3}} = C_{c,\text{mot}} = 1.3 \text{ kgm}^2$$

and its maximum joint torques are

$$M_{\text{mot},1} = M_{\text{mot},2} = M_{\text{mot},3} = M_{\text{mot}} = 10 \text{ Nm}.$$

For this task the manipulator's joint velocities and accelerations are required to be zero at the beginning and at the end of the trajectory, i.e.

$$\dot{q}_i\,(t = 0) = 0, \quad \dot{q}_i\,(t = t_{\text{E}}) = 0, \quad \ddot{q}_i\,(t = 0) = 0, \quad \ddot{q}_i\,(t = t_{\text{E}}) = 0, \tag{6.6}$$

where $i \in \{1, 2, 3\}$. This example is similar to the one in [9].

Application of inverse kinematics-based optimization approach

In the following, exemplary results obtained from the trajectory optimization methods derived in Section 4.3, which were applied on the example of the planar manipulator will be presented.

It was found that direct application of the methods presented in Section 4.3 was not ideally suited for this task as the convergence of the optimization problem was slow and the number of iterations exceeded a limit of 150 active-set iterations of the fmincon function of MATLAB R2014A in most cases. The method from Section 4.3 was then altered such that the trajectory end-time t_{E} was no longer a variable of the trajectory optimization problem but was subject to a higher-level optimization procedure. For a setup with trajectories of the path parameter s and the null space contribution $\ddot{q}_0$ with its scaling factor w the vector of optimization variables yields

$$\mathbf{x} = \left(\begin{array}{cc} \mathbf{d}_s^{\mathsf{T}} & \mathbf{d}_w^{\mathsf{T}} \end{array} \right)^{\mathsf{T}},$$

refer to (4.24) on page 42 for a comparison. In addition, the cost function was changed such that it incorporates the final joint velocities, i.e.

$$Z(\mathbf{x}) = \dot{\mathbf{q}}^{\top}(t_{\mathrm{E}}) \, \dot{\mathbf{q}}(t_{\mathrm{E}}).$$

By applying this modified method, convergence could be achieved in less than 100 iterations and the final joint velocities were negligible.

Figure 6.13 shows the normalized joint torques during the trajectory for the path with $\varphi = 180°$ illustrated in Figure 6.12 on page 68 that was obtained by applying the second-order inverse kinematics optimization method with directional kinematic manipulability extension, see Sections 4.3.2 to 4.3.7. The trajectory was computed for an end time of $t_{\mathrm{E}} = 2.6$ s, 46 control points and 50 uniformly distributed knots each for the trajectories of degree $n = 3$ of the path parameter s and the scaling factor w_{kin} for the null space contribution of the task space gradient of the directional kinematic manipulability. Results of the requirement (6.6) on the previous page are that $\dot{s}(t=0) = 0$ and $\dot{s}(t=t_{\mathrm{E}}) = 0$, hence only 40 of the 46 control points were utilized as optimization variables. The initial values for the control points $\mathbf{d}_s$ were chosen according to (5.2) on page 50. $\mathbf{d}_w$ was initialized with zeros. The optimization problem is constrained by joint torque limits at 1000 equally spaced points in time during the trajectory. From the fact that the joint torques are not used to full capacity at all times, see Figure 6.13, it is obvious that no minimum-time trajectory was achieved. Improvement can be made by further reducing the trajectory end-time t_{E} while adjusting the optimization setup.

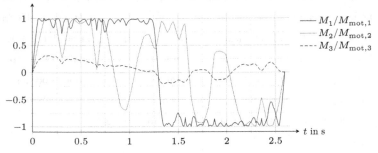

Figure 6.13: Planar robot — Inverse kinematics approach with directed kinematic manipulability, $\varphi = 180°$, results for normalized joint torques

Since the computation time for the above problem lies within the region of tens of hours using an INTEL XEON E3-1246 V3 processor, no further improvements to above solution were made. Solutions similar to Figure 6.13 on the previous page were obtained for many settings of φ in Figure 6.12 on page 68, but none of them completely satisfied the requirements.

Application of the separation-based optimization approach

In this section, it will be shown that the separation method presented in Chapter 5 can be applied in order to obtain minimum-time joint trajectories for the present example of the set of predefined geometric task space paths for the planar robot's end-effector according to Figure 6.12 on page 68.

As joint torque constraints are imposed onto the problem, it is required to compute the feed-forward torques for comparison. In order to avoid discontinuities of the joint torques, the degrees of the B-spline curve trajectories for the path parameter s and the angle of the *redundant* joint q_r were chosen to be $n_s = n_\mathrm{r} = 3$. The selection of higher degrees is possible but not advisable due to increased computation times for spline functions of higher degrees. The number of control points for each trajectory was chosen to be 46 from which 6 are not part of the optimization problem for meeting the requirement (6.6) on page 69. At 1000 equally spaced points in time during the trajectory with end time t_E the feed-forward torques were computed and compared against the physical limits as non-linear constraints to the optimization problem.

For the straight line path with angle $\alpha = 135°$ three trajectories with a very similar end time of $t_\mathrm{E} \approx 1.84$ s were obtained using different choices for the *redundant* degree of freedom q_r. Differences in the principal characteristics of the joint angles q_i apart from the exact end time are non-existent since the robot's initial pose is defined and the joint configuration does not change during the trajectory due to the specific form of the path without any singular configurations. Hence, only the graph for $q_\mathrm{r} = q_1$ is shown, see Figure 6.14 on the next page.

A striking result that can be seen from the visualization of the normalized joint torques

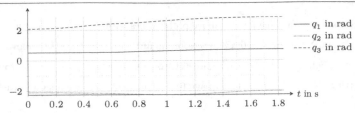

Figure 6.14: Planar robot — Separation approach, $q_{\mathrm{r}} = q_1$, $\varphi = 135°$, results for joint angles

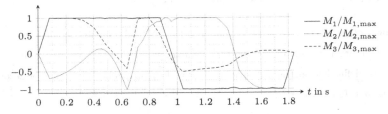

Figure 6.15: Planar robot — Separation approach, $q_{\mathrm{r}} = q_1$, $\varphi = 135°$, results for normalized joint torques

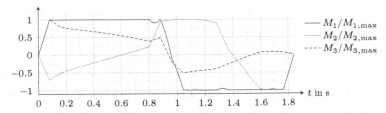

Figure 6.16: Planar robot — Separation approach, $q_{\mathrm{r}} = q_2$, $\varphi = 135°$, results for normalized joint torques

in Figure 6.15 to 6.17 is that the joint torques reach their limits most of the time during the trajectory. Comparing the trajectories of q_3 it can be observed that the selection of a joint as the *redundant* degree of freedom does not yield a tendency of the respective actuator to make full use of its torque limit.

Further reduction of the end time can be achieved by improving the local adaptiveness of both trajectories by increasing the number of control points for both trajectories in the optimization problem at the cost of increased computation times.

Optimization results for minimum-time trajectories for other choices of φ from Figure 6.12

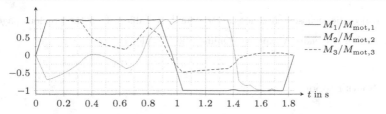

Figure 6.17: Planar robot — Separation approach, $q_r = q_3$, $\varphi = 135°$, results for normalized joint torques

on page 68 can be obtained from Appendix A (Springer Online Plus download).

Comparison of results

Table 6.1 lists a comparison of the end time of trajectories for this example achieved by means of different approaches, i.e. the approach described in [9] and the method featured in the present thesis. However, not all results may be directly compared since in [9] the results were obtained using the selection $q_r = q_3$. The symbol $\cdot$ is used for cases in which no convergence was achieved by means of the setup presented in the previous section.

Table 6.1: Planar example — Comparison of optimized trajectory end times in s

Subtask	φ	[9]	$q_r = q_1$	$q_r = q_2$	$q_r = q_3$
1	0°	1.79	1.87	1.88	$\cdot$
2	45°	2.00	2.10	2.11	2.29
3	90°	2.17	2.37	2.40	2.51
4	135°	2.35	1.84	1.84	1.84
5	180°	1.91	$\cdot$	$\cdot$	2.43
6	235°	1.84	$\cdot$	1.99	1.99
7	270°	1.55	1.59	1.94	1.59
8	315°	1.49	1.45	$\cdot$	1.45

Expectedly, the results from [9] yield shorter trajectory end times t_E in most cases since the level of continuity is lower than in the approach in the present thesis, i.e. the torques are discontinuous and therefore it is not required to reverse the torque direction. Subtracting the time intervals in which the joint torques change their active limits yields trajectory end times that are very similar to the results from [9].

The trajectories whose end times can be found in Table 6.1 on the previous page are visualized in Appendix A (Springer Online Plus download).

6.2 Spatial robot

This section will investigate the possibility of applying the method presented in this thesis on the task of finding a minimum-time trajectory for a STÄUBLI TX90L industrial robot with six degrees of freedom mounted on a linear axis.

First, the forward kinematics will be derived from the robot's geometric properties, then the separation method presented in Chapter 5 will be applied such that the linear axis degree of freedom is treated as *redundant* which allows to compute the well-known solutions for the remaining joints, i.e. for a 6 DOF serial robot with a spherical wrist. Finally, the optimization results obtained for a set of specified geometric paths are presented.

6.2.1 Forward kinematics

Figure 6.18 on the next page depicts the STÄUBLI TX90L industrial robot in a pose in which all of its joint actuators are in their initial position, i.e. $q_i = 0$. Using basic geometry, the robot's forward kinematics, i.e. the transformation from joint space coordinates q_i to the task space end-effector position r_E and the end-effector orientation, represented by the rotation matrix A_{IE} between the inertial frame and the end-effector coordinate frame, are derived. The numeric values of the physical dimensions can be found in Table 6.2.

Table 6.2: Dimensions of industrial robot

dimension	l_{Ix}	l_{Iy}	l_{Iz}	l_{Lz}	l_{1x}	l_{1y}	l_{1z}
value in m	1.507	0.181	0.098	0.268	0.05	0.140	0.210
dimension	l_{2x}	l_{2z}	l_{3x}	l_{3z}	l_{4z}	l_{5x}	
value in m	0.5	0.048	0.137	0.138	0.413	0.1	

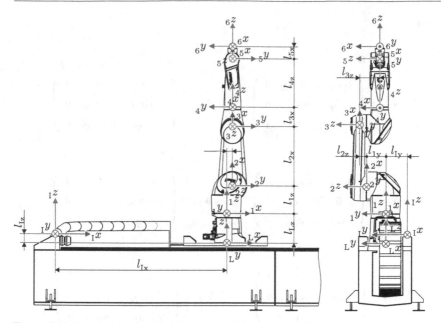

Figure 6.18: Industrial robot in initial pose with coordinate frames

End-effector orientation

While the end-effector orientation can be represented in different ways such as EULER angles or quaternions, the following derivations for the end-effector position in task space in the next section and the inverse kinematics solution in Section 6.2.2 will make use of a rotation matrix representation.

The rotation matrix $\mathbf{A}_{\mathrm{IE}}$, that represents the end-effector orientation, consists of the (non-commutative) composition

$$\mathbf{A}_{\mathrm{IE}} = \mathbf{A}_{\mathrm{IL}}\mathbf{A}_{\mathrm{L1}}\mathbf{A}_{12}\mathbf{A}_{23}\mathbf{A}_{34}\mathbf{A}_{45}\mathbf{A}_{56}$$

where

$$\mathbf{A}_{IL} = \mathbf{E}$$

$$\mathbf{A}_{L1} = \begin{pmatrix} \cos q_2 & -\sin q_2 & 0 \\ \sin q_2 & \cos q_2 & 0 \\ 0 & 0 & 1 \end{pmatrix} \qquad \mathbf{A}_{34} = \begin{pmatrix} 0 & 0 & 1 \\ -\sin q_5 & -\cos q_5 & 0 \\ \cos q_5 & -\sin q_5 & 0 \end{pmatrix}$$

$$\mathbf{A}_{12} = \begin{pmatrix} \sin q_3 & \cos q_3 & 0 \\ 0 & 0 & 1 \\ \cos q_3 & -\sin q_3 & 0 \end{pmatrix} \qquad \mathbf{A}_{45} = \begin{pmatrix} 0 & 0 & 1 \\ -\sin q_6 & -\cos q_6 & 0 \\ \cos q_6 & -\sin q_6 & 0 \end{pmatrix}$$

$$\mathbf{A}_{23} = \begin{pmatrix} \cos q_4 & -\sin q_4 & 0 \\ \sin q_4 & \cos q_4 & 0 \\ 0 & 0 & 1 \end{pmatrix} \qquad \mathbf{A}_{56} = \begin{pmatrix} 0 & 0 & 1 \\ -\sin q_7 & -\cos q_7 & 0 \\ \cos q_7 & -\sin q_7 & 0 \end{pmatrix}.$$

End-effector position

After obtaining the rotation matrices as described in the previous section, one may find the vector of end-effector coordinates in inertial representation $\mathbf{r}_E$ to be

$$_I\mathbf{r}_E = {}_I\mathbf{r}_{IL} + \mathbf{A}_{IL} \left({}_L\mathbf{r}_{L1} + \mathbf{A}_{L1} \left({}_1\mathbf{r}_{12} + \mathbf{A}_{12} \left({}_2\mathbf{r}_{23} + \mathbf{A}_{23} \left({}_3\mathbf{r}_{34} + \mathbf{A}_{34} \left({}_4\mathbf{r}_{45} + \mathbf{A}_{45} \, {}_5\mathbf{r}_{56} \right) \right) \right) \right) \right)$$

where

$$_I\mathbf{r}_{IL} = \begin{pmatrix} l_{Ix} + q_1 \\ 0 \\ 0 \end{pmatrix}$$

$$_L\mathbf{r}_{L1} = \begin{pmatrix} 0 \\ 0 \\ l_{Lz} \end{pmatrix} \qquad\qquad _3\mathbf{r}_{34} = \begin{pmatrix} l_{3x} \\ 0 \\ -l_{3z} \end{pmatrix}$$

$$_1\mathbf{r}_{12} = \begin{pmatrix} l_{1x} \\ l_{1y} \\ l_{1z} \end{pmatrix} \qquad\qquad _4\mathbf{r}_{45} = \begin{pmatrix} 0 \\ 0 \\ l_{4z} \end{pmatrix}$$

$$_2\mathbf{r}_{23} = \begin{pmatrix} l_{2x} \\ 0 \\ l_{2z} \end{pmatrix} \qquad\qquad _5\mathbf{r}_{56} = \begin{pmatrix} l_{5x} \\ 0 \\ 0 \end{pmatrix}.$$

6.2.2 Inverse kinematics $q_r = q_1$

For this type of industrial robot the linear axis with its position q_1 is considered the joint that may be most useful as the redundant coordinate for the separation approach presented in this document. In this section, the position of the linear axis joint q_1 is considered given and thus the inverse kinematics of the remaining standard six degrees-of-freedom manipulator with spherical wrist can be computed. For this robot configuration, in total eight different inverse kinematics solutions can be found because multiple physical joint configurations are possible to obtain the same end-effector position and orientation. These solutions will be below indicated by $\pm_i$ where $i \in \{1, 2, 3\}$.

Since the desired orientation of the end-effector $\mathbf{A}_{IE}$ is known, the position of the wrist joint can be computed, i.e.

$$_I\mathbf{r}_{I5} = {}_I\mathbf{r}_{I6} - \mathbf{A}_{IE6}\mathbf{r}_{56}$$

where $_6\mathbf{r}_{56} = \mathbf{A}_{65}{}_5\mathbf{r}_{56}$ and $\mathbf{A}_{IE} = \mathbf{A}_{I6}$. With the transformation

$$_L\mathbf{r}_{L5} = \mathbf{A}_{LI}\left({}_I\mathbf{r}_{I5} - {}_I\mathbf{r}_{IL}\right)$$

which includes only a translation because the axes of the I-system and the L-system are aligned, the translation

$$_L\mathbf{r}_{15} = \begin{pmatrix} _Lx_{15} \\ _Ly_{15} \\ _Lz_{15} \end{pmatrix} = {_L\mathbf{r}_{L5}} - {_L\mathbf{r}_{L1}}$$

and with the auxiliary angles

$$\alpha = \arctan\left(_Ly_{15}, {_Lx_{15}}\right)$$

$$\beta = \arcsin\left(\frac{l_{1y} + l_{2z} - l_{3z}}{\sqrt{_Lx_{15}^2 + {_Ly_{15}^2}}}\right)$$

the first rotation joint angle q_2 can be computed, i.e.

$$q_2 = \alpha \pm_1 \beta.$$

With the transformation

$$_1\mathbf{r}_{15} = \begin{pmatrix} _1x_{15} \\ _1y_{15} \\ _1z_{15} \end{pmatrix} = \mathbf{A}_{1L}{_L\mathbf{r}_{15}},$$

the auxiliary length

$$l = \sqrt{\left(_1x_{15} - l_{1x}\right)^2 + \left(_1z_{15} - l_{1z}\right)^2}$$

and the auxiliary angles

$$\gamma = \arctan\left(_1z_{15} - l_{1z}, {_1x_{15} - l_{1x}}\right) \text{ and}$$

$$\kappa = \arccos\left(\frac{l_{2x}^2 + l^2 - (l_{3x} + l_{4z})^2}{2\,l_{2x}\,l}\right),$$

the shoulder angle

$$q_3 = \frac{\pi}{2} - \gamma \pm_2 \kappa$$

and the elbow angle

$$q_4 = \arctan\left({}_1x_{15} - l_{1x} - l_{2x}\sin q_3, {}_1z_{15} - l_{1z} - l_{2x}\cos q_3\right)$$

can be found.

Now that the joint angles of the base, q_2, q_3 and q_4, are known, a numerical expression for the rotation matrix $\mathbf{A}_{3E}$ can be computed,

$$\mathbf{A}_{3E} = \mathbf{A}_{3I}\mathbf{A}_{IE} = \begin{pmatrix} a_{11} & a_{12} & a_{13} \\ a_{21} & a_{22} & a_{23} \\ a_{31} & a_{32} & a_{33} \end{pmatrix} \tag{6.7}$$

where the real numbers a_{ij} represent the entries of the matrix.

Equating (6.7) with the analytical expression for the same rotation matrix,

$$\mathbf{A}_{3E} = \begin{pmatrix} \sin q_6 \sin q_7 & \sin q_6 \cos q_7 & \cos q_6 \\ -\sin q_5 \cos q_7 - \cos q_5 \cos q_6 \sin q_7 & \sin q_5 \sin q_7 - \cos q_5 \cos q_6 \cos q_7 & \cos q_5 \sin q_6 \\ -\sin q_5 \cos q_6 \sin q_7 + \cos q_5 \cos q_7 & -\sin q_5 \cos q_6 \cos q_7 - \cos q_5 \sin q_7 & \sin q_5 \sin q_6 \end{pmatrix},$$

yields the remaining joint angles,

$$\text{Elements } (1,1) \text{ and } (1,2): \quad q_7 = \arctan(a_{11}, a_{12}) - s_3\pi$$
$$\text{Element } (1,3): \quad q_6 = \pm_3 \arccos(a_{13})$$
$$\text{Elements } (3,3) \text{ and } (2,3): \quad q_5 = \arctan(a_{33}, a_{23}) + s_3\pi$$

with the solution switch variable $s_3 = -\frac{\pm_3 1 - 1}{2}$.

Since the inverse kinematics expressions presented in this section only yield numeric results within certain bounds, the implementation must preserve joint position continuity in case of continuous task space trajectories by adding or subtracting 2π to a joint angle result if a discontinuity is detected.

The inverse kinematics solution for a similar 6 DOF industrial robot can be found in [1].

6.2.3   Optimization results

Task definition

The harmonized international standard EN ISO 9283:1998 [4] defines performance criteria of industrial robots such as path accuracy and repeatibility as well as related test methods. Amongst other tasks, paths are specified that are desired to be followed with constant velocity while measuring the path deviation.

To select an appropriate task for the application of the method presented in this thesis, a rectangular path from [4] was chosen. As recommended in the standard, the corners of the rectangle are rounded to avoid sudden changes in the end-effector velocity. Clothoides are used to obtain corner roundings that provide continuous slope and curvature and thus enabling joint accelerations and end-effector accelerations to be continuous. Figure 6.19 depicts an example path. As shown in Figure 6.20 on the next page, the desired rectangular path is located in a test cube whose edges are aligned with the inertial coordinate frame of the robot. The plane containing the rectangular path was selected to be in a diagonal plane of the cube. While the position of the test cube relative to the inertial coordinate frame varies between the subtasks, the end-effector orientation will remain normal to the testing plane. The paths are chosen such that the linear axis of the robot can remain at $q_1 = 1.5$ m to follow the path. It will be investigated whether enabling the linear axis results in trajectories with shorter end times.

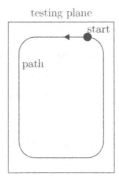

Figure 6.19: Example path — Rectangle rounded with clothoides

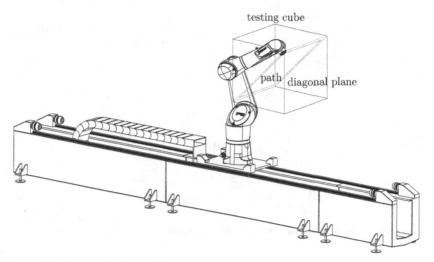

Figure 6.20: Industrial robot with rectangular path in testing cube

The edge length of the testing cube depicted in Figure 6.20 is chosen to be $a = 0.63$ m from the set of permitted sizes in [4]. As required, the dimensions of the rectangle are 80% of the corresponding edge lengths of the diagonal square in the testing cube. Since the standard does not explicitly require clothoides for rounding the rectangle's corners, the clothoides' parameters were freely chosen to be $\varphi_c = 0.1745$ rad and $r_c = \frac{a}{10} = 0.063$ m. In [8] the mathematical function describing clothoides is derived. Figure 6.21 on the next page shows one corner rounding

$$\mathbf{p}\left(s\right) = \begin{pmatrix} x\left(s\right) \\ y\left(s\right) \end{pmatrix}, \quad s \in [0, 1].$$

Figure 6.22 on page 83 depicts the slopes in the x and y directions, i.e.

$$x' = \frac{\mathrm{d}x}{\mathrm{d}s}, \quad y' = \frac{\mathrm{d}y}{\mathrm{d}s}$$

and the continuous curvature along $\mathbf{p}\left(s\right)$, i.e.

$$\kappa = \frac{x'y'' - x''y'}{\left(x'^2 + y'^2\right)^{3/2}} \quad \text{where} \quad x'' = \frac{\mathrm{d}^2x}{\mathrm{d}s^2}, y'' = \frac{\mathrm{d}^2y}{\mathrm{d}s^2}$$

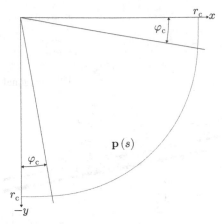

Figure 6.21: Corner rounding with clothoides, example

where the path parameter arguments s were omitted.

For each subtask, the start position of the end-effector on the path depicted in Figure 6.19 on page 80 is listed in Table 6.3. For subtasks 1 to 3 the end-effector orientation remains constant such that the $_6z$ axis is perpendicular to the plane with the path and pointing towards the ground whereas for subtasks 4 and 5 the $_6z$ axis points away from the ground.

Table 6.3: Start positions for subtasks

Subtask	$_Ix$ in m	$_Iy$ in m	$_Iz$ in m
1	2.5	0	0.7
2	2.5	0.63	0.7
3	2.85	0.63	0.7
4	3.15	0.63	0.8
5	3.5	0.22	0.87

Optimization parameters

The method that was used to conduct the optimization for minimal-time trajectories along the end-effector paths defined above can be looked up in Chapter 5 with the selection of the linear axis to be treated as the *redundant* degree of freedom, i.e. $q_r = q_1$.

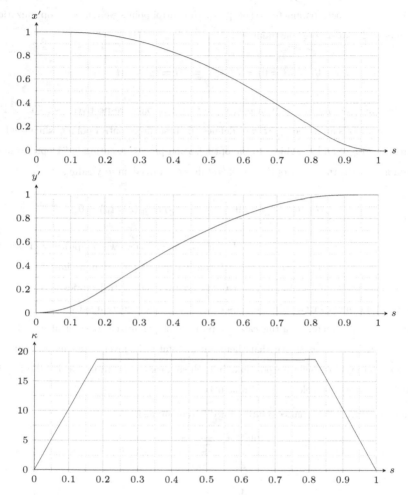

Figure 6.22: Corner rounding with clothoides, slopes x' and y', curvature κ

The maximum polynomial degree of the trajectories for the path parameter s and the *redundant* degree of freedom were chosen to be $n_s = n_r = 3$ because of the continuity needed for the computation of the feed-forward torques that are to be compared with the physical constraints. For both trajectories, a uniform distribution according to (2.2) on page 5 with 80 knots each was selected. Each trajectory is defined by means of 76 control points. For the path parameter $s(t) \in [0,1]$, 70 control points were used as optimization variables. To satisfy the constraint

$$s(t=0) = \dot{s}(t=0) = \ddot{s}(t=0) = \dot{s}(t=t_{\mathrm{E}}) = \ddot{s}(t=t_{\mathrm{E}}) = 0,$$

the first and last $n_s = 3$ control points are zeros, or ones, respectively, see (5.2). For $q_{\mathrm{r}}(t)$ the robot's initial pose $q_{\mathrm{r}}(t=0)$ defines the first three control points, while the last three are also equal but subject to the optimization process resulting in 71 optimization variables for the trajectory of the *redundant* degree of freedom q_{r} yielding

$$\dot{q}_{\mathrm{E}}(t=0) = \ddot{q}_{\mathrm{E}}(t=0) = \dot{q}_{\mathrm{E}}(t=t_{\mathrm{E}}) = \ddot{q}_{\mathrm{E}}(t=t_{\mathrm{E}}) = 0.$$

In the optimization process the initial value of the end time was empirically chosen to be $t_{\mathrm{E}} = 1.5$ s. In contrast to the method from Chapter 5 for obtaining initial values for the control points of $s(t)$ and $q_{\mathrm{r}}(t)$ where the resulting initial trajectories were found by means of a numeric inverse kinematics solution for a linear function between zero and one for the path parameter s, in this case the position of q_{r} remains constant at $q_{\mathrm{r}} = 1.5$ m. In Section 6.2.2 it was shown that there are in total eight inverse kinematics solutions possible for every end-effector position. In each subtask one solution that is valid for the whole trajectory was selected, see Table 6.4.

Table 6.4: Inverse kinematics solution for different subtasks

Subtask	$\pm_1$	$\pm_2$	$\pm_3$
1	+	+	+
2	−	−	+
3	+	+	+
4	−	−	+
5	−	−	+

The optimization was conducted by utilizing the Active Set algorithm of the `fmincon` function implemented in MATLAB R2014A. The physical limits of the robot that constrain the desired motion are listed in Table 6.5 on the next page. Following Section 5, those constraints are imposed on the optimization process via non-linear constraints that are checked at multiple times during the trajectory, in this case at 1000 equally spaced points in time.

Table 6.5: Physical limits of industrial robot

Joint i	Gear ratio	$M_{i,\max}$ ($\pm$)	$\ddot{q}_{i,\max}$ ($\pm$)	$\dot{q}_{i,\max}$ ($\pm$)	$q_{i,\min}$	$q_{i,\max}$
1	114.2727	69 Nm	15 m/s^2	4 m/s	0 m	2.55 m
2	32	42 Nm	770°/s^2	400°/s	$-180°$	180°
3	32	42 Nm	482°/s^2	400°/s	$-130°$	147.5°
4	32	17.5 Nm	1084°/s^2	400°/s	$-145°$	145°
5	32	4.5 Nm	3082°/s^2	500°/s	$-270°$	270°
6	45	3.4 Nm	2042°/s^2	450°/s	$-115°$	140°
7	30	2.2 Nm	6000°/s^2	720°/s	$-270°$	270°

Results

In this section, selected results of the optimization process described above are presented. As it can be seen from the results for all five subtasks in Appendix B (Springer Online Plus download), the characteristics are very similar. Obviously, the physical limits for the joint accelerations $\ddot{q}_i$ are chosen very conservatively resulting in the joint torques M_i not reaching their physical limits. The computation time for each subtask lies in the region of two hours on an INTEL XEON E3-1246 V3 processor using the current implementation in MATLAB R2014A.

It can be seen from the results that the joint acceleration constraints are violated for short time intervals on multiple occasions, which results from too large constraint checking intervals. The violations only occur between and not at the checking points in time. It is also clear that no trajectory with a strict minimum-time property was achieved since there exist short time intervals at which none of the limits are reached or exceeded.

Since the position of the linear axis q_1 was set to be constant for the initial trajectory, but the trajectory resulting from the optimization problem shows changes it follows that in this case trajectories with lower end times can be obtained by enabling the linear axis to move freely within its physical limits.

Improvements to the results can be made by reducing the constraint checking interval times which yields constraints that are more often satisfied. Increasing the number of control points and thus the number of knot intervals leads to better local adaptivity and thus lower end times. However, both suggestions come at the price of higher computational cost and thus longer computation times.

7. Conclusion

Kinematically redundant serial robots are increasingly gaining attention for industrial applications due to their extraordinary versatility and compliance especially in structured workspaces. As shorter cycle times yield improved cost-effectiveness of robotic operations, the requirement for time-optimimal trajectories along predefined task space paths arises. Because of topological differences to conventional non-redundant manipulators additional complexity is added to the task of trajectory planning.

While methods for point-to-point trajectory planning for kinematically redundant serial manipulators can be found in special literature, only a small number of approaches to finding trajectories along predfined geometric paths exists. Results obtained using methods that yield strict time-optimal trajectories are mostly not directly applicable to real manipulators as the continuity level of the resulting trajectories is low which can induce oscillations and may not be feasible due to the physical limits of the robot drives. Other approaches yield trajectories of arbitrary continuity but lack procedural efficiency.

This thesis presents two fundamentally different approaches to solve the problem of finding minimum-time trajectories for kinematically redundant manipulators along predefined task space paths. In the first method the trajectory of the parameter that parametrizes the task space path is described as a time-dependent B-spline curve. The trajectories of the individual robot joints are then computed by means of first and second-order inverse kinematics approaches. Additionally, those trajectories are improved by exploiting a kinematically redundant manipulator's capability of performing null-space motion. Time-optimality is pursued by means of an optimization problem in which the trajectory end time is minimized and which is inequality constrained by physical and technological limitations of the robot. The second procedure presented in this thesis is based on a separation approach known from literature in which a kinematically redundant robot's joints are devided in two sets. For one selected robot joint a time-dependent B-spline curve is assumed as its trajectory. A separate B-spline trajectory with the same end time is assumed for the task space path parameter. From the task space path trajectory and the trajectory of the separated joint it is now possible to determine the trajectories of the remaining joints by means of analytical inverse kinematics. Trajectories with minimum end time are a result of an optimization process with the control points of the trajectories of the path parameter and the separated joint as parameters. Physical and technological restrictions of the manipulator are incorporated as inequality constraints to the optimization problem.

The presented methods for obtaining minimum-time trajectories were applied to two different examples of kniematically redundant serial manipulators, a planar three-link SCARA and a spatial industrial robot with seven degrees of freedom. For the approach based on numerical inverse kinematics it was found that the convergence of the optimization problem is slow and thus the computation times exceed feasible limits on current hardware due to the formulation. A possible direction of future work may be to investigate whether reformulating the optimzation problem leads to increased efficiency. Applying the separation-based method yielded time-optimal trajectories with respect to the given constraints. In special cases it even outperformed an approach from literature that generally yields trajectories with shorter end time due to lower continuity requirements. Encouraging results were obtained that can be further improved by modifying certain parameters of the methodology. Another opportunity is the investigation of continuation methods for analytical inverse kinematics solutions enabling the separation-based algorithm to be applied to task space paths of increased complexity.

Bibliography

[1] Mathias Brandstötter, Arthur Angerer, and Michael Hofbaur. An analytical solution of the inverse kinematics problem of industrial serial manipulators with an ortho-parallel basis and a spherical wrist. In *Proceedings of the Austrian Robotics Workshop 2014*, 2014.

[2] Hartmut Bremer. *Elastic Multibody Dynamics: A Direct Ritz Approach.* Springer Verlag, 2008.

[3] T. Chettibi, H.E. Lehtihet, M. Haddad, and S. Hanchi. Minimum cost trajectory planning for industrial robots. *European Journal of Mechanics A/Solids*, 23:703–715, 2004.

[4] International Organization for Standardization. EN ISO 9283:1998, February 1998.

[5] Hubert Gattringer. *Starr-elastische Robotersysteme: Theorie und Anwendungen.* Springer Verlag, 2011.

[6] Knut Graichen. *Vorlesungsskriptum Optimierung.* Institut für Automatisierungs- und Regelungstechnik, TU Wien, 2009.

[7] B. Jüttler. *Skriptum zur Vorlesung Geometrische Methoden.* Institut für angewandte Geometrie, Johannes Kepler Universität Linz, 2013.

[8] Johannes Kilian. Quasistatischer Folgeregler und Bahnoptimierung für einen Segway. Master's thesis, Johannes Kepler Universität Linz, Institut für Robotik, 2009.

[9] Shugen Ma and Mitsuru Watanabe. Time optimal path-tracking control of kinematically redundant manipulators. JSME International Journal, 2004.

[10] Wayne Piegl, Les; Tiller. *The NURBS Book.* Monographs in visual communication. Springer Verlag, second edition edition, 1997.

[11] Kurt Schlacher. *Vorlesungsskriptum Automatisierungstechnik I.* Johannes Kepler Universität Linz, 2006.

[12] Bruno Siciliano and Oussama Khatib, editors. *Handbook of Robotics*. Springer Verlag, 2008.

[13] Bruno Siciliano, Lorenzo Sciavicco, Luigi Villani, and Giuseppe Oriolo. *Robotics - Modelling, Planning and Control*. Advanced Textbooks in Control and Signal Processing series. Springer, 2009.

[14] The MathWorks Inc. Constrained nonlinear optimization algorithms, September 2014.

Printed in the United States
By Bookmasters